A-C CIRCUIT ANALYSIS

Edited by
Alexander Schure, Ph.D., Ed.D.

JOHN F. RIDER PUBLISHER, INC.
116 West 14th Street • New York 11, N. Y.

PREFACE

In many design applications, alternating current offers substantial advantages over direct current. Among the more important advantages of alternating current are ease of transmission over long distances without considerable loss of power and its ability to be radiated as energy into space from an antenna. For these reasons, as well as the fact that alternating current may easily be converted to direct current, ac has become accepted as a more suitable and versatile power source than direct current. Although there are many situations in which direct current is a proper choice as a fundamental source of power (for example, in any mobile installation such as the automobile, where a d-c storage battery represents the initial source of power) there are correspondingly many more situations in which it becomes essential to utilize and understand alternating current.

The aim of this book is to provide the fundamental concepts of alternating current analysis. As in D-C CIRCUIT ANALYSIS (another text in this series) the mathematical treatment is simple, but the analyses are extensive enough to allow the interested technician or student to develop a full comprehension of the pertinent theory. To insure the achievement of this aim, the text presents adequate information relating to electrical laws in such form as to permit ready use; it describes a relatively small number of selected major topics in *detail,* rather than treating a major body of less important material; the topics, once given, are related to practical situations; carefully selected problems afford the reader more profitable information and an opportunity to apply the principles he has learned;

step-by-step problem analyses provide clearcut concepts of the methodology involved in solving problems.

Specific attention is given to basic concepts: simple a-c waveforms and terminology; complex a-c waveforms: a-c generation; two-pole and four-pole alternators; the radian; appropriate mathematics pertaining to the voltage and current values of a sine wave including average, effective and peak relationships; the concepts of "pure" resistance, inductance and capacitance; a complete treatment of the "j" operator, including complex numbers; the polar form of complex numbers; the effects and calculations arising from various combinations of R, L and C in series circuits; the series resistance circuit; the vectorial treatment used in analyses of parallel networks; the series-parallel a-c network; and, alternate methods of computing impedence. Thus, a strong theoretical basis is provided, upon which more advanced concepts can be built.

Grateful acknowledgement is made to the staff of the New York Institute of Technology for its assistance in the preparation of the manuscript of this book.

November, 1958   A. S.
New York, N. Y.

CONTENTS

ELECTRONIC TECHNOLOGY SERIES

GROUP 1: ELECTRICAL BASES FOR ELECTRONICS

Electrostatics
D-C Circuit Analysis
A-C Circuit Analysis
Magnetism*
Resonant Circuits
R-C/R-L Time Constant

GROUP 2: BASIC ELECTRONICS

Vacuum Tube Rectifiers
Vacuum Tubes
   Characteristics
Impedance Matching
Gas Tubes
Limiters & Clippers
Transistor Funda-
   mentals*

GROUP 3: AMPLIFICATION

Audio Amplifiers*
R-F Amplifiers*
Magnetic Amplifiers*
Video Amplifiers*
Grid Bias & Coupling
   Systems*
Inverse Feedback

GROUP 4: OSCILLATORS AND PROPAGATION

L-C Oscillators
Crystal Oscillators
Multivibrators
Blocking Oscillators
R-F Transmission Lines
Wave Propagation

GROUP 5: COMMUNICATION ELECTRONICS

Amplitude Modulation
A-M Detectors
Superheterodyne
   Converters & I-F
   Amplifiers
Frequency Modulation
FM Limiters & Detectors
Antennas

GROUP 6: ELECTRONIC ENGINEERING FUNDAMENTALS

Microwave Tubes*
Electron Optics*
Waveguides*
Television Engineering*
Radar Systems*
Computers*

* In Preparation

Chapter 1

BASIC PRINCIPLES OF ALTERNATING CURRENT

An alternating current is one that starts from zero amplitude, rises in a specified manner to some maximum value, then falls off to zero again. The current next reverses its direction of flow, rises to its maximum value in the opposite direction, then decays again to zero. Then there is another reversal of current direction and the cycle of variation occurs again.

1. Simple A-C Waveforms and Terminology

Figure 1 shows the changes described above, and illustrates the definition of fundamental terms used in ac. Figure 1A shows two cycles of a type of ac called a *sine wave*. This particular shape is the *graph* of the trigonometric function, the sine of an angle, as we shall show in detail later. An alternating current generated by rotating machinery is shaped this way; that's why the sine wave is the most important waveform we shall study.

A cycle is a complete set of variations of an alternating current. As marked off in Fig. 1, a cycle is shown from the beginning of one positive alternation to the beginning of the next positive alternation. Any other two corresponding points on the waveform could have been selected as well. The cycle is shown marked off into two half cycles, one for the positive alternation and one for the negative alternation.

The length of time consumed by one cycle is called the "period" of the ac. Period is symbolized by "t" and it is in units of time — seconds, microseconds, etc.

The number of cycles of ac occurring in 1 second is called its "frequency." The symbol for frequency is "f" and it is in units of cycles per second (cps). High frequencies may be stated in kilocycles (1000 cycles) per second or megacycles (1,000,000 cycles) per second. For simplicity, the phrase "per second" is often omitted, and frequency is stated in cycles, "c," kilocycles, "kc," or megacycles, "mc."

Since *period* is the length of time for one cycle (seconds per cycle), and frequency is the number of cycles in 1 second (cycles per second), then period and frequency are evidently reciprocals. We then have,

$$f = 1/t$$

where f is in cycles per second and

$$t = 1/f$$

t is in seconds.

Example 1. What is the period of a 60-cycle ac?

Solution. $t = 1/f = 1/60 = 0.0167$ sec.

Example 2. The period of an ac is 2.5 milliseconds. What is its frequency?

Solution. $f = 1/t = 1/.0025 = 400$ cycles

Referring to Fig. 1A again, note that the *peak* or *maximum* value of the ac is indicated. For this sine wave it is evidently 5 amperes.

2. A Complex A-C Waveform

Another type of waveform more complex than the sine wave is shown in Fig. 1B. Since the vertical axis is marked off in volts, this is an alternating *voltage* rather than an alternating *current*. For simplicity, the term ac is used to describe *both* alternating voltages and alternating currents. Actually it is an alternating voltage that is generated and the current that flows depends on the voltage and the characteristics of the load.

To fully appreciate the importance of ac, we must realize that about 95% of the electrical power generated in the U.S. is a-c power. Ac is far superior to the d-c systems once used because it is easy to generate, can be transmitted over long distances with low losses, and facilitates the increase or decrease of voltage with relatively simple and highly efficient devices (transformers).

3. Electromagnetic Induction — A-C Generation

Now, to better understand the nature of alternating voltages and currents, we will discuss the generation of ac by rotating machinery. The principles of electromagnetic induction as discovered in 1831 by Michael Faraday are applied in this process.

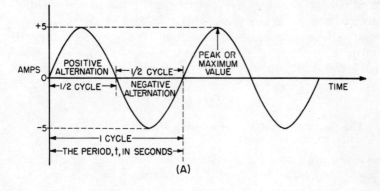

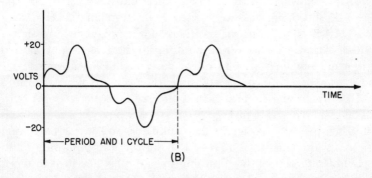

Fig. 1.   A-C waveforms: (A) sine-wave alternating current; (B) a complex alternating voltage.

The electric generator which produces ac by electromagnetic induction is called an "alternator."

Faraday demonstrated that a voltage is induced in a conductor whenever the conductor cuts magnetic lines of force. It is imma-

terial whether the conductor or the lines of force are in motion
(even both may be moving). Relative motion between the two
resulting in a cutting of magnetic flux lines is all that is required
for the production of a voltage. Faraday showed further that the
polarity of the induced voltage depended on the direction of the
magnetic field and the direction of cutting the lines. A reversal of
either the field or the relative motion can produce a reversal of
polarity of the induced voltage.

Finally, Faraday demonstrated that the *magnitude* of the induced voltage is in direct proportion to the rate of cutting of the
lines of force. This last statement is the basis of the definition of
a *volt*. If the magnetic field is uniform throughout, and the relative speed of the conductor cutting the magnetic flux is constant,
then 1 volt is generated when the conductor cuts 10^8 (100,000,000)
lines of force per second.

4. The Simple Two-Pole Alternator

These principles may be utilized in constructing a simple alternator. The left side of Fig. 2 illustrates a basic two-pole (two magnetic poles) alternator. Conventionally, the direction of the lines
of force is shown from the north to the south pole. A single conductor is shown in cross section revolving through the field at a
constant velocity. Thirteen positions of the conductor are shown
with an arrow at each position pointing to the direction of motion
at that moment. Position 13 is obviously identical with position 1
and marks the end of one revolution. The conductor revolves in a
path shown by the circle in which the positions lie.

The right-hand side of Fig. 2 is a graph whose vertical (y) axis
represents induced voltage in the conductor, and whose horizontal
(x) axis represents time. However, if we consider position 1 as an
angle of 0°, then the other positions may also be described in terms
of the angle swept out by the conductor. The positions shown rep-
resent angular distances 30° apart. In this case, the horizontal axis
may also be marked out in degrees as is shown.

We now proceed to take the conductor through the various positions in one revolution. The amplitude of the induced voltage
will be marked off on the graph in accordance with the principle
of rate of cutting lines of force.

In position 1, the conductor is moving in a direction parallel to the lines of force. In this case it is not cutting any force lines and the induced voltage is zero as shown. In going from position 1 to 2 it is evident that the direction of motion is constantly changing such that more and more lines will be cut. The result is a slowly rising voltage from 1 to 2. Position 2 represents an angular traverse

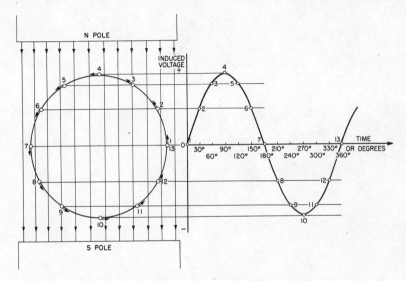

Fig. 2.   Generation of a-c voltage by a revolving conductor in a uniform magnetic field

of 30°. From position 2 to 3 at 60° an even greater number of lines of force will be cut in unit time; the voltage curve continues its rise.

In position 4 at 90°, the direction of motion is at right angles to the force lines and maximum voltage is induced at this point. By trigonometric considerations it can be shown that the number of lines cut by the conductor (hence the magnitude of the induced voltage) is proportional to the sine of the angle corresponding to the position of the conductor. At 30°, the induced voltage is thus 0.5 of the maximum at 90°. At 60°, the voltage is 0.866 of the maximum.

Positions 5 and 6 are analogous to 3 and 2 respectively, with the difference that the voltage is now decreasing. At 7, the conductor is once again moving parallel to the lines of force and no voltage is induced (trigonometrically, sin 180° = 0).

As we go from 7 to 8 the situation is the same as in going from 1 to 2, with one fundamental difference. From 1 to 2, the conductor moved from right to left in cutting the field. From 7 to 8, the motion of the conductor is from left to right through the field. In accordance with the previous discussion, the polarity of the induced

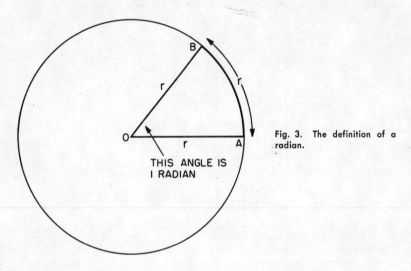

Fig. 3. The definition of a radian.

voltage must now reverse. As a result the voltage goes to a negative maximum at 10 and decreases to zero again at 13. At 13 one cycle is completed and the next revolution repeats the induced voltage waveforms.

The graph of the induced voltage is identical with the curve y = sin x. For this reason the output voltage of the alternator is called a sine wave. We have already indicated this in the fact that the number of lines cut by the conductor in unit time depends on the sine of the angle.

5. The Radian

Up to this point, angles have been expressed in the customary manner, in degrees. However, angles may be expressed in terms of another unit, the radian. The radian measure of angles often leads to a better understanding of a-c concepts.

To define a radian, we draw a circle of radius r, as in Fig. 3. On the circumference of the circle we lay out an arc AB equal in length to the radius. Draw in the two radii, OA and OB. By definition, angle BOA is *one radian*.

By a basic geometric formula, the whole circumference of the circle is given by,

$$C = 2\pi r$$

This means that the arc, AB, which is equal to r, can be put around the circumference 2π times. Since each arc subtends (forms) one radian, then there must be 2π radians in the whole circle.

A circle then contains 2π radians. In ordinary measure a circle consists of 360°. In that case,

$$2\pi \text{ radians} = 360°$$

and

$$\pi \text{ radians} = 180°$$

from which

$$1 \text{ radian} = \frac{180°}{\pi} = 57.3°$$

and

$$1° = \frac{\pi}{180°} = 0.0175 \text{ radian}$$

Example 3. Express the following in radians. (a) 21°, (b) 60°, (c) 240°.

Solution.

(a) $21° = 21 \times 0.0175 = 0.368$ radian

(b) $60° = 60 \times \frac{\pi}{180} = \frac{\pi}{3}$ radians

(c) $240° = 240 \times \frac{\pi}{180} = \frac{4\pi}{3}$ radians

Example 4. Express the following in degrees.

(a) 4 radians, (b) $\frac{2\pi}{3}$ radians, (c) $\frac{7\pi}{6}$ radians

Solution.

(a) $4 \text{ radians} = 4 \times 57.3° = 229°$

(b) $\frac{2\pi}{3} \text{ radians} = \frac{2\pi}{3} \times \frac{180°}{\pi} = 120°$

(c) $\frac{7\pi}{6} \text{ radians} = \frac{7\pi}{6} \times \frac{180°}{\pi} = 210°$

In the basic two-pole alternator of Fig. 2, one revolution of the conductor around the magnetic field develops one cycle of induced alternating voltage. In the radian system of angular measurement, a circle consists of 2π radians. If we speak of the velocity of the conductor in terms of *angular velocity* (radians per second) rather than conventional linear velocity (such as inches per second) a useful relationship can be developed.

Let ω (omega) be the angular velocity of the conductor in radians per second. Since one cycle of induced voltage is produced for each 2π radians, then the frequency of the voltage is

$$f = \omega/2\pi$$

and

$$\omega = 2\pi f$$

The angular velocity, ω, is an important and recurring concept in ac theory; but it is important to remember that this concept employs radians and not degrees.

6. The Four-Pole Alternator

In practice, the simple two-pole alternator is not an efficient machine. Actual alternators are multipole types. Electromagnets

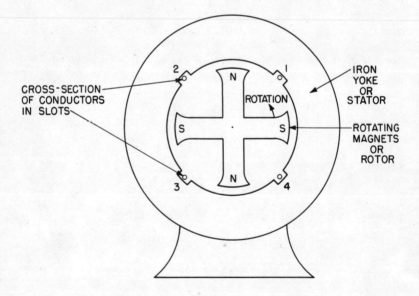

Fig. 4.   A four-pole alternator.

are used to produce the magnetic field and a d-c voltage must be supplied to generate the magnetic poles. Since this "field voltage" is a low dc it is best to make the poles the moving element or the *rotor* of the alternator. The d-c voltage may then be brought in through low voltage slip rings and brushes. The conductors wherein the voltage is generated form a winding called the *armature*. Since it is the stationary part of the alternator, the armature is often called the *stator*.

Figure 4 is a simplified drawing of a four-pole rotating field alternator. For additional simplicity only four armature conductors are shown. The iron frame labeled "yoke" is for mechanical support and also completes the magnetic circuit.

Since the lines of force go from a north pole to the adjacent south poles, it is evident that one cycle will be completed from one north pole under conductor 1 (or any conductor) to the next north pole under that conductor. For one rotor rotation, two cycles of voltage are induced. Had there been six poles, (the number of poles, of course, must be even) the ratio would be three cycles for one rotation.

The frequency of the induced ac is thus seen to be a function of the number of poles and the rate of rotation of the rotor. In equation form,

$$f = \frac{p}{2} \times rps$$

where f is frequency in cycles per second, p is total number of poles, rps is revolutions per second.

Since machine speeds are customarily expressed in revolutions per minute, rpm, the frequency equation becomes

$$f = \frac{p}{2} \times \frac{rpm}{60}$$

and

$$f = \frac{p\,(rpm)}{120}$$

Example 5. An eight-pole alternator is rotating at a speed of 900 rpm. What is the frequency of the induced voltage?

Solution. $f = \dfrac{p\,(rpm)}{120} = \dfrac{8\,(900)}{120} = 60$ cycles

Example 6. At what speed should a 20-pole alternator rotate to generate a frequency of 400 cycles?

Solution. Solving the frequency equation for rpm,

$$rpm = \frac{120f}{p} = \frac{(120)\,(400)}{20} = 2400$$

7. Review Questions

(1)   Explain the action of one cycle of ac.

(2)   What is the period of a 100-cycle ac?

(3)   The period of an a-c voltage is 150 μsec. What is its frequency?

(4)   What is the relationship between voltage, lines of force, and angular velocity?

(5)   An alternator has two major components. Which one is moving? Why?

(6)   The voltage output of an alternator is equivalent to which trigonometric function?

(7)   An alternator generates a maximum of 20 volts. What is its instantaneous voltage at 30°? At 60°?

(8)   Convert 300° into π radians.

(9)   What is the equation for frequency for a multipole alternator?

(10) Define angular velocity.

Chapter 2

VOLTAGE AND CURRENT VALUES OF A SINE WAVE

In a d-c power system we can speak of the voltage and the current as fixed numbers and perform Ohm's law and power calculations with these numbers. In an a-c system, however, voltage and current are constantly changing in sine-wave fashion. It thus becomes necessary to define them on other bases than the simple d-c basis.

8. The A-C Sine-Wave

Figure 5 shows a sine-wave of alternating voltage. The voltage is zero at $0°$, $180°$ and $360°$. It reaches its positive maximum at $90°$ and its negative maximum at $270°$. Between these angles, the voltage is evidently at intermediate values between zero and maximum.

Let us call the absolute value of the maximum voltage E_m.* This is also referred to as the peak voltage. We say the "absolute" value so that we can eliminate the plus sign. E_m is thus treated as a positive number unless we specifically wish to refer to the voltage at $270°$.

The voltage at any instant in time (or at any angle) is called the "instantaneous" value of the voltage. The symbol, E_i is used for instantaneous voltage. Comparably, an alternating current will have I_m for maximum current and I_i for instantaneous current.

Since an a-c sine wave is the graph of the function $y = \sin x$, we can relate maximum and instantaneous values by the sine of the

---

* E is the conventional symbol for voltage, I for current.

angle corresponding to the instant under consideration. The Greek letter theta, Θ, is used as a symbol for the angle. On this basis we have the following relationships,

$$E_i = E_m \sin \Theta$$

and, $$E_m = E_i/\sin \Theta$$

Correspondingly, $$I_i = I_m \sin \Theta$$

$$I_m = I_i/\sin \Theta$$

In solving problems involving these formulas, we must remember the properties of the sine in each of the four quadrants. They are summarized here:

Quadrant I $(0°-90°)$
   a) The sine is a positive number.
   b) $\sin \Theta$ read directly from a trig table.*

Quadrant II $(90°-180°)$
   a) The sine is a positive number.
   b) $\sin \Theta = \sin (180° - \Theta)$

Quadrant III $(180°-270°)$
   a) The sine is a negative number.
   b) $\sin \Theta = \sin (180° + \Theta)$

Quadrant IV $(270°-360°)$
   a) The sine is a negative number.
   b) $\sin \Theta = \sin (360° - \Theta)$

Example 7.   The maximum value of an alternating voltage is 145 volts. What is the instantaneous voltage at 45° of its cycle?

Solution. $$E_i = E_m \sin \Theta$$

$$= 145 \sin 45° = 145 \times 0.707$$

$$= 102 \text{ v}$$

Example 8.   The maximum value of an ac is to be found. The instantaneous value at 165° is 7.25 amps. Find I_m.

Solution. $$I_m = I_i/\sin \Theta = 7.25/\sin 165°$$

$$\sin 165° = \sin (180° - 165°) = \sin 15° = 0.259$$

$$I_m = 7.25/0.259 = 28.0 \text{ amperes}$$

---

* A Trigonometry Table is included at the end of this volume.

Example 9. The maximum value of an ac is 15.5 amps. At what points in the cycle will the current be −11.0 amps?

Solution. An examination of a sine wave shows that it goes through any given value other than zero and its maximum twice in a cycle. Two angles must be found for this solution. Solving the relationship for sin Θ, we get

$$\sin \Theta = I_1/I_m$$

$$= -11.0/15.5 = -0.709$$

The angle in the table with a sine closest to the absolute value of 0.709 is 45°, whose sine is 0.707. However, since this sine is negative we must find the equivalent third and fourth quadrant angles. The third quadrant angle is

$$\Theta_1 = 180° + 45° = 225°$$

while the fourth quadrant solution is

$$\Theta_2 = 360° - 45° = 315°$$

Note that in these examples the solutions are calculated to 3 significant figures. This is the standard accuracy of a 10-inch slide rule and no more than slide-rule accuracy is generally required in the solution of practical problems. This assumes, of course, that the given data is correct to three figures, an assumption that will be made throughout this volume unless otherwise specified. In general, a calculated result should have no more significant figures than the smallest number of significant figures in the least accurate item of its data.

9. The Average Value of an A-C Sine-Wave

Another useful value to be considered for an a-c sine wave is its "average" value. The average value of a full *cycle* of a sine wave

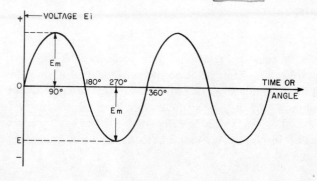

Fig. 5. A sine-wave.

is evidently zero as the negative alternation exactly cancels the positive alternation. Therefore, when the term "average value" is applied to an ac it is understood to mean the average of one

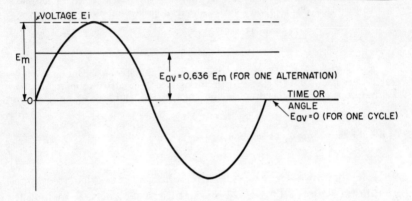

Fig. 6. Average values of a sine-wave.

alternation. Since only absolute values are of concern, the average value will be represented by a positive number.

E_{av} and I_{av} are the symbols used for average voltage and average current, respectively. By mathematical analysis we can show that these relationships hold true,

$$E_{av} = \frac{2}{\pi} \times E_m = 0.636 \ E_m$$

and

$$E_m = \frac{\pi}{2} \times E_{av} = 1.57 \ E_{av}$$

Similarly,

$$I_{av} = 0.636 \ I_m$$

$$I_m = 1.57 \ I_{av}$$

The average value relationship of 0.636 can be shown approximately by taking enough instantaneous values in one-half cycle and finding their average. This is done in the table below which is based on a half cycle of a sine wave with a peak value of 10 volts. The instantaneous value is calculated for angles 5° apart up to 180°. The instantaneous value is in the column headed E_i and is calculated by

$$E_i = 10 \ \sin \Theta$$

TABLE I
INSTANTANEOUS
VOLTAGE AND CURRENT VALUES

Angle (degrees)	E_i (volts)	I_i^2* (amps)	Angle (degrees)	E_i (volts)	I_i^2* (amps)
5	0.87	0.8	95	9.96	99.2
10	1.74	3.0	100	9.85	97.0
15	2.59	6.7	105	9.66	93.3
20	3.42	11.6	110	9.40	88.3
25	4.23	17.9	115	9.06	82.1
30	5.00	25.0	120	8.66	75.0
35	5.74	33.0	125	8.19	67.1
40	6.43	41.4	130	7.66	58.8
45	7.07	50.0	135	7.07	50.0
50	7.66	58.8	140	6.43	41.4
55	8.19	67.1	145	5.74	33.0
60	8.66	75.0	150	5.00	25.0
65	9.06	82.1	155	4.23	17.9
70	9.40	88.3	160	3.42	11.6
75	9.66	93.3	165	2.59	6.7
80	9.85	97.0	170	1.74	3.0
85	9.96	99.2	175	0.87	0.8
90	10.00	100.0	180	0.00	0.0

If we now add the 36 values of E_i the total turns out to be 229.06. Dividing by 36 we get an average value of 6.36 for the 10-volt peak wave. This checks the 0.636 relationship between E_{av} and E_m. Figure 6 shows the relation between E_{av} and E_m.

Example 10. What is the average value of a voltage whose peak value is 188 volts?

Solution. $$E_{av} = 0.636 \ E_m = 0.636 \times 188$$
$$= 120 \text{ volts}$$

Example 11. An alternating current has an average value of 1.53 amps. What is the peak value of the current?

Solution. $$I_m = 1.57 \ I_{av} = 1.57 \times 1.53$$
$$= 2.40 \text{ amps}$$

* Column I_i^2 is not being used until a later discussion.

10. The Effective (rms) Value of an A-C Sine-Wave

We have now discussed the peak, instantaneous and average values of an ac. However, in order to carry out calculations involving power and heating, it is necessary to introduce a new concept, the "effective" value of an a-c wave. (This is also called the "root mean square" or rms value.) The effective value is best derived from a discussion of power in an a-c circuit.

The instantaneous power in ac is defined as the product of the instantaneous voltage and current. Thus,

$$P_i = E_i I_i$$

The average of all the instantaneous powers in a cycle is the average power consumed by the load. The positive and negative half cycles contribute equally to the power consumption; as far as heat produced and power consumed are concerned, it is immaterial in which direction the current flows through a resistor.

Let us give the average power the symbol, P. By methods of integral calculus it is possible to determine the *average* power in an a-c circuit *containing resistance only* as

$$P = \frac{E_m I_m}{2}$$

where E_m and I_m are the peak values. This power equation may be rewritten as

$$P = \frac{E_m}{\sqrt{2}} \times \frac{I_m}{\sqrt{2}} = (0.707 \ E_m) \ (0.707 \ I_m)$$

Using the symbols E and I for *effective* voltage and current, we define effective values from this power relationship.

$$E = 0.707 \ E_m$$

from which $E_m = E/0.707 = 1.41 \ E$

and $I = 0.707 \ I_m$

from which $I_m = 1.41 \ I$

The power equation for an a-c resistive circuit may now be written simply as

$$P = EI = I^2 R$$

This is identical with the d-c power equations. The effective values of voltage and current are thus equivalent to the d-c voltage and current values that develop the same amount of power in a resistor.

11. Effective Value Relationships

The 0.707 relationship between effective and peak values may also be approximated arithmetically. Since power is related to the square of the current, if we average enough squares of instantaneous currents we will get the 0.707 relationship.

Let us go back to the table that was used for the development of average values. The third column in Table I, headed $I_i{}^2$, gives the value of the square of the instantaneous current for each of the angles. This is based on a peak current of 10 amperes.

Adding up the 36 values of $I_i{}^2$ gives a total of 1800.4. Dividing this total by 36 gives the effective current (I) squared, *i.e.* 50. The square root of 50 (7.07) is the effective current. This is for a 10-amp peak current and thus bears out the 0.707 relationship.

The above derivation shows that the number 0.707 comes about as a result of squaring currents, taking the average or *mean* value of the squares and then extracting the square root. This explains why the effective value of voltage or current is also called the "root mean square," a name that simply describes the process of obtaining this value. Root mean square may be abbreviated to *rms*. Effective and rms values are synonymous.

Figure 7 shows a sine wave of voltage with peak, average and rms values. The same is true for a current wave.

The effective or rms value is the one most generally used. When we speak of so many volts or amps we mean effective voltage or current, unless otherwise stated. Voltmeters and ammeters are normally calibrated so that they read *effective* values. Note, however, that the product of rms voltage with rms current is *average* power.

Relations exist between average and effective values as follows,

$$E = 0.707 \ E_m = 0.707 \ (1.57 \ E_{av})$$

therefore, $$E = 1.11 \ E_{av}$$

and $$E_{av} = E/1.11 = 0.9 \ E$$

$$I = 1.11 \ I_{av}$$

$$I_{av} = 0.9 \ I$$

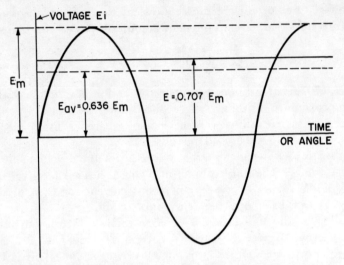

Fig. 7.   Peak, effective (rms), and average values of a voltage sine-wave.

Example 12. In a resistive a-c circuit a voltmeter across a 250-ohm resistor reads 120 volts. Find rms, maximum and average current, maximum and average voltage and the power dissipated.

Solution. Since the voltmeter reads rms volts, we may find rms current directly from Ohm's law, which states that current is proportional to voltage and inversely proportional to resistance.

$$I = E/R = 120/250 = 0.480 \text{ amp}$$

$$I_m = 1.41 \ I = 1.41 \times 0.480 = 0.676 \text{ amp}$$

$$I_{av} = 0.9 \ I = 0.9 \times 0.480 = 0.432 \text{ amp}$$

$$E_m = 1.41 \ E = 1.41 \times 120 = 169 \text{ volts}$$

$$E_{av} = 0.9 \ E = 0.9 \times 120 = 108 \text{ volts}$$

$$P = EI = 120 \times 0.480 = 57.5 \text{ watts}$$

We may check the power with this equation:

$$P = I^2R = (0.480)^2 \times 250$$

$$= 57.5 \text{ watts}$$

12.   Review Questions

(1)   Express instantaneous voltage (E_i) in terms of maximum voltage (E_m).
(2)   Express I_m in terms of I_i in a resistive a-c circuit.
(3)   A generator delivers a maximum of 200 volts. What is the instantaneous value at 330°?

(4) Express average current (I_{av}) in terms of maximum current (I_m).

(5) What is the average voltage of the generator in Question 3?

(6) Express maximum voltage (E_m) in terms of effective voltage (E_{rms}).

(7) A generator has an output of 14.4 volts maximum. What is the power dissipation of a 20-ohm resistor connected across its output?

(8) A 100-ohm resistor is connected across a 220-volt a-c line. Find I_{rms}, I_m, I_{av}, E_m and power consumption.

(9) In which quadrants is the sine a negative number?

(10) Express P_{av} in terms of E_m and I_m.

Chapter 3

RESISTANCE, INDUCTANCE AND CAPACITANCE

In the study of d-c circuits, we learned that the only limit to the current amplitude in the steady-state condition was the resistance of the circuit. It was only during the short periods following the closing or the opening of the circuit that inductance or capacitance had any effect. These periods constitute the *transient condition* of the circuit.

In a-c circuits the voltage and current are constantly varying. As a result, inductance and capacitance exert a profound effect on the current. The combination of L, C, and R with various other devices such as tubes, transistors, etc., make up the multitudinous electrical and electronic circuits used in power work, communications, industrial controls, and so on. In this chapter we will study the individual effects of resistance, inductance and capacitance when an alternating voltage is applied to each.

13. "Pure" Resistance

A practical resistor inevitably contains some small amount of inductance in its leads or in its turns of wire (if it is wire wound). There is also some stray capacitance inherent in its construction. However, for analytical purposes, let us define a "pure" resistor as a device that has no inductance or capacitance and may be represented by the letter R. We will employ this useful fiction again for L and C.

If a pure resistance is placed across a source of alternating voltage, an alternating current will flow whose amplitude is determined by Ohm's law. Using *effective* values of voltage and current, we have the same relationship as exists in a d-c circuit.

$$I = E/R$$

The instantaneous currents through the resistor always follow the instantaneous variations of the voltage as it goes through its cycle. This may be stated in another way. In Fig. 8 the curve marked E represents the applied voltage. The current curve I follows the voltage variations exactly. We will define such a relationship by saying that E and I are "in phase." They are identical sine-waves differing only in amplitude.

Since P = EI for a pure resistive circuit we can sketch in the curve for P in Fig. 8. Note first that P is always positive, since the product of two positive numbers or two negative numbers is a positive number. Physically, a positive power is power delivered from the source and consumed by the load. In this case the power develops heat in the resistor.

A second interesting conclusion drawn from the power curve is the fact that for each cycle of E or I there are *two* cycles of power. The frequency of the power variations is *twice* the frequency of the voltage variations. For a 60-cycle voltage the power frequency is 120 cycles. Note that the power is zero twice in every voltage cycle, a fact that must be reckoned with in the design of single-phase a-c motors. This major disadvantage of single-phase a-c led to the development of polyphase systems.

14.   "Pure" Inductance

For the analysis of an inductive circuit we will again use the fiction of a "pure" inductance containing no resistance or capacitance. Such a device is physically impossible, but useful for study purposes. The symbol "L" stands for pure inductance.

Any piece of wire, even a straight piece, contains some inductance. Generally, however, we think of inductance in terms of a coil consisting of several turns of wire around an air core or magnetic core.

Consider the situation where an alternating current is passed through a coil. This current sets up a magnetic field which follows exactly the variation of the current. The magnetic field then is an alternating one corresponding to the alternating current. This magnetic field, by its very nature, is constantly moving as it rises, falls, reverses, rises and falls. In so moving, it is constantly cutting the conductors which make up the coil, thus (by Faraday's laws) *inducing* a voltage in the coil.

By the physical laws of inertia, this induced voltage must be of such a polarity as to *oppose* the changing current which is producing the magnetic field. It is evident that if this induced voltage

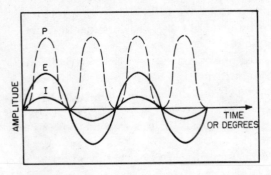

Fig. 8. P, E, and I in a purely resistive circuit.

aided the changing current, the magnitude and rate of change of the current would increase. This, in turn, would increase the size of the induced voltage which, in turn, would further increase the changing current. This snowballing process is analogous to a perpetual motion situation and is physically impossible.

The direction of the induced voltage is summarized by Lenz' law which states that *the induced voltage is of such a polarity as always to oppose the change of the current which produces it.* Because of its direction this voltage is often called the countervoltage. The amplitude of the countervoltage is proportional to the *rate of change* of the current through the coil.

This property of generating a countervoltage is called inductance. The type of inductance discussed with the coil is often called "self inductance" since only one coil is involved, and all the effects take place within it.

The countervoltage forms the basis of the definition of the unit of inductance, the *henry*. A device has an inductance of 1 henry when it generates a countervoltage of 1 volt when the current through it is changing at the rate of 1 ampere per second.

The property of inductance in electricity is analogous to the property of inertia in mechanics. Both inductance and inertia tend to maintain the *status quo* in opposition to any *change* in the forces acting upon them. For this reason, inductance is sometimes referred to as *electrical inertia*.

Several interesting effects turn up in a circuit containing inductance only. The power dissipated in such a circuit is zero. $P = I^2R$, and since $R = 0$, then $P = 0$. This is true regardless of the magnitude of the current, and, in a-c circuits which are highly inductive, leads to the situation where large currents flow with very little power expenditure. This is evidently an unhappy situation for electric companies whose business is to sell electrical power.

A second important effect of inductance in a-c is the condition of the steady state of an a-c shown in Fig. 9. This occurs sometime after the circuit has been energized and all transient effects have vanished.

The curve marked I is the sine wave of current passing through L. At $0°$ the current is zero, but its *rate of change is* maximum.

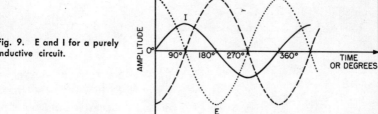

Fig. 9. E and I for a purely inductive circuit.

(At zero the current is going from nothing to something, which is certainly a great rate of change.) Since the rate of change of I is maximum, then the countervoltage must be maximum. Further, since the current is trying to rise in the positive direction, the countervoltage in opposing this change must be negative. The dashed line curve of the figure marked E_{ind} represents this induced countervoltage. At $0°$ it is at its negative maximum.

From $0°$ to $90°$ the rate of change of I decreases and E_{ind} decreases. At the $90°$ point, I is at its maximum value, but its rate of change is zero. It is at "the top of the hill," poised to come down. The magnetic field around the inductance is at a maximum, but it is stationary. As a result, the countervoltage is zero.

From 90° to 180° the rate of change of I increases with a corresponding increase of E_{ind}. However, the current is decreasing and the induced voltage must be positive to oppose the current change. By similar analysis the rest of the curve of E_{ind} may be drawn.

The force which is driving the current to go through the various changes just described is the *applied* voltage. Thus, we have a situation where the applied voltage *produces* the current changes and the countervoltage *opposes* the current changes. As a result, the countervoltage and the applied voltage must be 180° out of phase with each other.

The applied voltage, E, is thus drawn in on Fig. 9 as the dotted line curve. It is a sine wave 180° out of phase with the countervoltage, E_{ind}.

Now let us compare the sine wave E, the voltage applied to the inductance, with the sine wave I, the current flowing through the inductance. E passes through its positive maximum at 0°, while I has its positive maximum at 90°. E goes through zero at 90°, while I reaches zero at 180°. Further comparisons show that I is displaced from E by 90°. Since I reaches its various values 90° later than E, I is said to *lag* E by 90°.

This angular difference between the voltage and current waves is called the phase angle. For a pure inductive circuit we express the phase angle as 90° lagging.

In Fig. 10, E and I are redrawn using E as the reference wave. If the power is drawn in from the relation P = EI it will be the curve marked P. P is negative when either E or I is negative, and positive when both have the same sign. Note that the P wave occurs with twice the frequency of E and I (just as in the resistive circuit). However this time it is an alternating wave with equal positive and negative half cycles. As a result, the *average* power over a cycle is zero. During the positive alternation the source delivers power to the inductance, but during the negative alternation the inductance returns the power to the source.

We are most accustomed to working with positive power which is defined as power delivered from the source to the load. This occurs when E and I are both positive or both negative. Negative power is power delivered from the load to the source. This occurs during the periods when E is positive and I negative, *or* E is negative and I positive. E and I must have opposite signs for a period of negative power to occur.

As a physical interpretation, during periods of positive power, the magnetic field around the inductance builds up. During periods of negative power the field collapses returning the stored-up energy to the source.

One effect of the induced or countervoltage *in an inductor* is the lagging current previously described. Another important effect is its limiting action on the *amplitude* of current flow. The counter-voltage acts to cancel out most of the applied voltage and thus acts to reduce the circuit current.

This limiting action on current amplitude is best described in terms of ohms rather than volts to make it analogous to the limiting action of a resistance on the current. We thus define the term

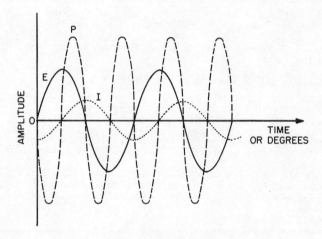

Fig. 10.   P, E and I in a purely inductive circuit.

inductive reactance as the *property of an inductor that limits the amplitude of an alternating current.* The symbol for inductive re-actance is X_L and its unit is the ohm. An inductance has a reactance of 1 ohm when 1 volt of applied voltage causes a current of 1 ampere to flow. Ohm's law for an inductance may now be written as

$$E = IX_L$$

The reactance of an inductor is determined by two factors. The size of the inductance will evidently determine X_L (a larger L will produce a larger countervoltage). Since the countervoltage also depends on rate of change of current, it is evident that a higher-frequency current with its more rapid fluctuations will produce a higher X_L than a lower-frequency current. It is the angular velocity, ω $(2\pi f)$, that determines the rate of current change. As a result, the following relationship holds,

$$X_L = 2\pi fL = \omega L$$

In this equation, X_L is in ohms, f in cycles per second and L is in henries.

Example 15. A 200-millihenry coil in a 60-cycle circuit has what inductive reactance?

Solution.
$$X_L = 2\pi fL$$
$$= 6.28 \times 60 \times 0.200$$
$$= 75.4 \text{ ohms}$$

Example 16. At what frequency will a 250-microhenry coil have a reactance of 235 ohms?

Solution. Solving the reactance formula for f,

$$f = \frac{X_L}{2\pi L} = \frac{235}{6.28 \times 250 \times 10^{-6}}$$
$$= 0.150 \times 10^6$$
$$= 150 \text{ kc}$$

Example 17. In a pure inductive circuit, the applied voltage is 150 volts at 400 cycles. The current is measured at 2.6 amps. What is the inductance of the circuit?

Solution. By Ohm's law

$$X_L = E/I = 150/2.6 = 57.7 \text{ ohms}$$

Solving the reactance equation for L,

$$L = \frac{X_L}{2\pi f} = \frac{57.7}{6.28 \times 400} = 0.0230 \text{ hy}$$
$$= 23 \text{ millihenries}$$

We will now examine the effect of placing inductances in series with each other. Assume that three coils are placed in series in such a way that there is no magnetic interaction between individual coils (Fig. 11). Magnetic interaction between coils is called mutual inductance. The rules developed for three series inductances may then be extended for any number.

Each coil exerts a limiting effect on the current which we have named inductive reactance. The effect of the three coils then, is a total limiting effect which is the sum of the individual reactances. Call the total inductive reactance X_{LT}, then

$$X_{LT} = X_{L1} + X_{L2} + X_{L3}$$

Rewriting,

$$\omega L_T = \omega L_1 + \omega L_2 + \omega L_3$$

Dividing through by ω,

$$L_T = L_1 + L_2 + L_3$$

The total inductance for inductances in series is then simply the sum of the individual inductances, when there is no mutual inductance. This is analogous to the total resistance of series resistances.

The case of inductances in parallel, as shown in Fig. 12, is different. Again, assuming that the mutual inductance among the coils

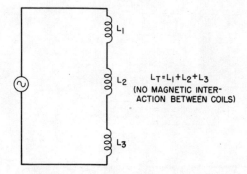

Fig. 11. Series inductances with no mutual inductance.

$L_T = L_1 + L_2 + L_3$
(NO MAGNETIC INTER-
ACTION BETWEEN COILS)

is zero, for a three-branch parallel network, the total current is expressed as the sum of the individual branch currents.

$$I_T = I_1 + I_2 + I_3$$

If each branch is an inductance, then by Ohm's law the current equation may be written as,

$$\frac{E}{X_{LT}} = \frac{E}{X_{L1}} + \frac{E}{X_{L2}} + \frac{E}{X_{L3}}$$

Dividing through by E and replacing X_L by ω_L,

$$\frac{1}{\omega L_T} = \frac{1}{\omega L_1} + \frac{1}{\omega L_2} + \frac{1}{\omega L_3}$$

Multiplying through by ω,

$$\frac{1}{L_T} = \frac{1}{L_1} + \frac{1}{L_2} + \frac{1}{L_3}$$

This reciprocal formula is analogous to the familiar formula for resistances in parallel.

15. "Pure" Capacitance

For the analysis of the effect of capacitance in an a-c circuit we also use the concept of a pure capacitor containing no inductance or resistance.

A capacitor consists of two conductors separated by a nonconductor called a dielectric. When a voltage is placed across a capacitor there is an instant rush of current which charges the capacitor to

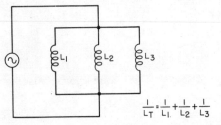

Fig. 12. Parallel inductances with no mutual inductance.

$$\frac{1}{L_T} = \frac{1}{L_1} + \frac{1}{L_2} + \frac{1}{L_3}$$

(NO MAGNETIC INTERACTION BETWEEN COILS)

the applied voltage. The current then stops since the voltage across the capacitor is equal and opposite to the applied voltage. If the voltage should now tend to increase, there will be an additional current flow in the same direction building up the counter-voltage on the capacitor until it is once more equal to the applied voltage.

If the voltage should tend to decrease, the capacitor loses some of its charge, causing a current flow in the *opposite* direction. In an a-c circuit where there is a continual change in the applied voltage, a capacitor is continually charging and discharging. While there is no current flow *through* the capacitor, the continual charging and discharging constitute a current flow in the rest of the circuit.

It is evident from the above discussion that the effect of capacitance is to *oppose a change in voltage* in an a-c circuit. This contrasts with inductance which opposes a change in *current*. The greater the rate of change of the voltage the more rapidly will the capacitor charge or discharge to oppose the change. The voltage and current relations that arise from this are shown in Fig. 13.

Figure 13 illustrates the steady-state condition some time after the capacitive circuit is energized and transient effects are over. The applied voltage E is shown as a sine wave. Any change in E is immediately opposed by a corresponding change in the countervoltage across the capacitor. As a result the countervoltage must be 180° out of phase with E. It is shown as the dashed-line curve marked E_c.

Now consider that the countervoltage is developed by current flowing to charge the capacitor. The greater the rate of change of the applied voltage the greater must be this charging current flow to enable the charge on the capacitor to catch up with the applied voltage.

The greatest rate of change of E occurs at 0° of its cycle. As a result I must be at a maximum at this point. From 0° to 90° the rate of change of E decreases and the charging current correspond-

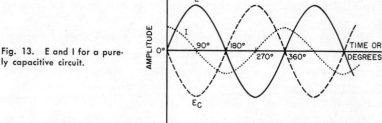

Fig. 13.   E and I for a pure-ly capacitive circuit.

ingly decreases. At 90° the rate of change of E is zero and the current flow in the circuit is also zero.

After the 90° point, the applied voltage starts to fall off at an increasing rate of change. The capacitor now discharges back into the source causing a reversal of the direction of current flow. With an increasing rate of change of E the current rises in amplitude reaching a maximum at 180°. The reversal of the source voltage now continues the current in the same direction, but with a decreasing rate of change of E the current decreases. At 270° the rate of change of E becomes zero and the current is zero. With the falling off of the source voltage the capacitor again discharges and the current reverses direction.

The dotted line marked I in Fig. 13 shows the current wave. An inspection of the curves for E and I shows that I goes through its alternations 90° ahead of E. In a pure capacitive circuit I *leads* E by 90°. The phase angle is thus 90° leading.

Figure 14 shows the relationship of voltage, current and power in the capacitive circuit. Again the power frequency is twice the

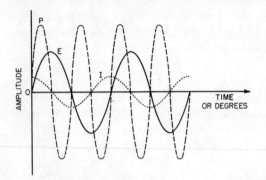

Fig. 14. P, E and I in a purely capacitive circuit.

applied frequency and the average power is zero, since the positive and negative alternations are equal.

Here again we meet the concept of positive and negative power. We first encountered this in the discussion of power in an inductive circuit. Again positive power is considered to be power going from the source to the load, while negative power is taken as power returned from the load to the source.

Periods of positive power occur when E and I have the same sign (both + or both −). Periods of negative power occur when E and I have opposite signs (one is + when the other is −).

During the periods of positive power the capacitor is charging and takes power from the source to build up its electric field. During the periods of negative power the capacitor is discharging and the energy in its electric field is returned to the source.

The unit of capacitance is the farad. A capacitor has a capacitance of one farad when a voltage across it of one volt charges it with one coulomb of electricity.* A farad is an enormous unit, and

---

* See *D-C Circuit Analysis*, A. Schure, (ed.), (1958: John F. Rider, Publisher, Inc.)

more practical units are the microfarad (10^{-6} farad) and the micro-microfarad (10^{-12} farad).

In addition to producing a lead of 90° of the current with respect to the voltage, a capacitor also acts to limit the amplitude of the current. Since the countervoltage is always in opposition to the applied voltage it acts to limit the current flow. This current limiting action of a capacitor is called its *capacitive reactance*. This is expressed in ohms and symbolized as X_C.

X_C is analogous to the X_L of an inductance and the R of a resistance, and Ohm's law applies to a capacitive circuit. Expressed by Ohm's law, the current through a purely capacitive circuit is

$$I = E/X_C$$

Just as with an inductance, there are two factors that determine the reactance of a capacitor. The *larger* the value of C, the *larger* will be the charging current necessary to build up the countervoltage at any given rate of change of applied voltage. A *larger* current means a *smaller* X_C. X_C then is inversely proportional to C.

A *higher* frequency means a *higher* angular velocity and hence a *larger* rate of change of the current. A *larger* rate of change of the current implies a *larger* current flow to build up the countervoltage. A *larger* current again means a *smaller* X_C. X_C then is also inversely proportional to the angular velocity. This may be expressed in equation form as follows,

$$X_C = \frac{1}{\omega C} = \frac{1}{2\pi fC}$$

where X_C is in ohms, f is in cycles per second and C is in farads.

For purpose of computation, $\frac{1}{2\pi}$ may be evaluated as 0.159, and the formula then written as

$$X_C = \frac{0.159}{fC}$$

Example 18. A 0.33-microfarad capacitor is placed across a 250-volt, 400-cycle source. What current will flow?

Solution.

$$X_C = \frac{0.159}{fC} = \frac{0.159}{400 \times 0.33 \times 10^{-6}} = 1210 \text{ ohms}$$

$$I = E/X_C = 250/1210$$

$$= 207 \text{ milliamps}$$

Example 19. What size capacitor is required to give a reactance of 765 ohms at a frequency of 10.4 megacycles?

Solution.

$$C = \frac{0.159}{fX_C} = \frac{0.159}{10.4 \times 10^6 \times 765}$$

$$= 20 \times 10^{-12} \text{ farads}$$

$$= 20 \text{ micromicrofarads}$$

If three capacitors, C_1, C_2 and C_3 were placed in series, as in Fig. 15, each would contribute a limiting effect on the current equal to

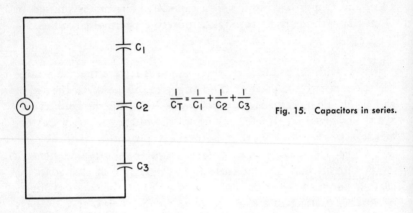

$$\frac{1}{C_T} = \frac{1}{C_1} + \frac{1}{C_2} + \frac{1}{C_3}$$

Fig. 15. Capacitors in series.

its reactance. The total limiting, or total capacitive reactance, X_{CT}, would then be the sum of the individual reactances, or

$$X_{CT} = X_{C1} + X_{C2} + X_{C3}$$

Rewriting,

$$\frac{1}{\omega C_T} = \frac{1}{\omega C_1} + \frac{1}{\omega C_2} + \frac{1}{\omega C_3}$$

Multiplying both sides of the equation by ω,

$$\frac{1}{C_T} = \frac{1}{C_1} + \frac{1}{C_2} + \frac{1}{C_3}$$

From this equation it is evident that capacitors in *series* add, just as resistors in parallel. The reciprocal rule applies.

For the case of capacitors in parallel (Fig. 16), we first solve for total current. For three capacitive branches,

$$I_T = I_1 + I_2 + I_3$$

Replacing each current by its Ohm's law equivalent,

$$\frac{E}{X_{CT}} = \frac{E}{X_{C1}} + \frac{E}{X_{C2}} + \frac{E}{X_{C3}}$$

Dividing both sides of the equation by E,

$$\frac{1}{X_{CT}} = \frac{1}{X_{C1}} + \frac{1}{X_{C2}} + \frac{1}{X_{C3}}$$

But,

$$\frac{1}{X_{CT}} = \frac{1}{\dfrac{1}{\omega C_T}} = \omega C_T$$

Each of the right-hand terms may be similarly transformed to give,

$$\omega C_T = \omega C_1 + \omega C_2 + \omega C_3$$

Dividing both sides by ω, we get the desired expression,

$$C_T = C_1 + C_2 + C_3$$

Capacitors in parallel add like resistors in series. The total capacitance is the sum of the individual capacitances.

Example 20. Three capacitors of the following values, 20 microfarads, 40 microfarads and 10 microfarads, are placed in series. What is the total capacitance?

Solution.

$$\frac{1}{C_T} = \frac{1}{C_1} + \frac{1}{C_2} + \frac{1}{C_3} = \frac{1}{20} + \frac{1}{40} + \frac{1}{10} = \frac{2}{40} + \frac{1}{40} + \frac{4}{40} = \frac{7}{40}$$

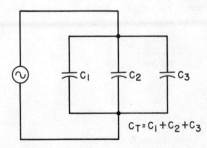

Fig. 16. Capacitors in parallel.

Taking reciprocals,

$$C_T = \frac{40}{7} = 5.72 \text{ microfarads}$$

Note that the total capacitance of series capacitors must be less than the smallest individual capacitor. This is the same as the case with resistors in parallel.

Example 21. What is the total parallel capacitance of the three capacitors in the previous example?

Solution.

$$C_T = C_1 + C_2 + C_3$$

$$= 20 + 40 + 10 = 70 \text{ microfarads.}$$

In a parallel connection, the total capacitance must be greater than the greatest individual capacitance. This is the same as in the case of resistors in series.

16. Review Questions

(1)   Define the relationship existing between voltage and current in a circuit containing
   a. resistance only;
   b. capacitance only;
   c. inductance only.

(2)   What is the reactance of a 2-henry choke at 400 cps?

(3)   A capacitor has a reactance of 200 ohms at 100 kc. What is its capacitance?

(4)   What determines the amplitude of the countervoltage in an inductance?

(5)   What is the power consumption of a circuit with inductance only?

(6)   Two inductances of 6 henries each are in parallel across a circuit. What is their effective inductance?

(7)   Two capacitors of 6 microfarads each are in parallel across a circuit. What is their combined capacitance?

(8)   Magnetic interaction between coils is known by what name?

(9)   Define 1 henry.

(10) Define 1 farad.

Chapter 4

THE J OPERATOR

17. The "Imaginary" Number

In mathematics, the quantity $\sqrt{-1}$ arises in the solution of numerous problems. There is no real number which when squared is equal to -1, hence there is no real number which is equal to $\sqrt{-1}$. In mathematics this quantity is termed an "imaginary" number, and is assigned the symbol, i.

It happens that $\sqrt{-1}$ is an important and useful concept in a-c theory and its use simplifies the handling and solution of many problems. However, to avoid the confusion of "i" with the symbol for current, we employ the symbol j. We then make our basic definition,

$$j = \sqrt{-1}$$

In electrical work, j is employed as an "operator." We define multiplication by j as equivalent to the operation of a 90° rotation. That this is reasonable and leads to consistent results is demonstrated in Fig. 17. Take any positive real number such as $+4$ in the figure. Multiplying by j rotates the number by 90° to the position marked $+j4$. (Note the convention of writing j *before* the number.) The vertical line upward from the origin is established as the $+j$ axis.

Another multiplication by j rotates the number by an additional 90°, bringing it to the -4 position. Two operations by j have effectively multiplied the original number by (-1). This is consistent since two multiplications by j is the same as multiplication by j ×j. But,

$$j \times j = j^2 = (\sqrt{-1})^2 = -1$$

A third multiplication by j produces another 90° rotation. The original number has now been rotated by 270° and is at the $-j4$

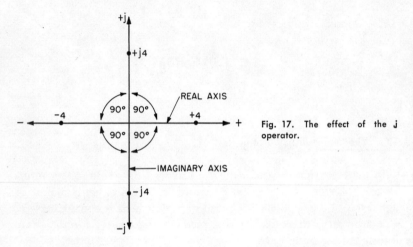

Fig. 17. The effect of the j operator.

position. The downward vertical line is called the $-j$ axis. This too is consistent since we have now multiplied by j × j × j, and

$$j \times j \times j = j^3 = j^2 \times j = (-1) \times j = -j$$

Finally, a fourth multiplication by j makes a complete 360° rotation and the number has returned to the original $+4$ position. Arithmetically,

$$j \times j \times j \times j = j^4 = j^2 \times j^2 = (-1)(-1) = +1$$

To summarize, the j operator has the following arithmetic and geometric meanings:

$$j = \sqrt{-1} \qquad \text{implies a } 90° \text{ rotation}$$
$$j^2 = -1 \qquad \text{implies a } 180° \text{ rotation}$$
$$j^3 = -j \qquad \text{implies a } 270° \text{ rotation}$$
$$j^4 = 1 \qquad \text{implies a } 360° \text{ rotation}$$

Higher powers of j simply mean new cycles of rotation. For example, $j^5 = j$, $j^6 = j^2 = -1$, etc.

The horizontal axis in Fig. 17, marked + to the right and − to the left of the origin, is called the *real* axis. On this axis lie all the real numbers of arithmetic from minus infinity to plus infinity. The vertical axis, marked + j upward and − j downward from the origin, is called the *imaginary* axis. All "pure" imaginary numbers, defined as the even roots of negative numbers, lie on this axis.

18. Complex Numbers

Any number lying on any part of the plane except the axes is called a "complex" number. Such a number has two parts or components, a real component and an imaginary component. Figure 18

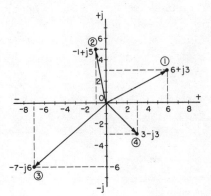

Fig. 18. Complex numbers.

shows an example of a complex number and the conventional way of writing such numbers.

Point 1 is the complex number, 6 + j3. This implies that to locate this number we must find a point on the complex plane + 6 to the right of the origin and + 3 up on the + j axis. Similarly, point 2 is − 1 + j5. This is one unit to the left of the origin and 5 units upward. Point 3 is − 7 − j6, and point 4 is 3 − j3 by similar reasoning.

This form of notation, of writing a complex number as the sum of its real and imaginary components, is called the "rectangular form" of the complex number. The general representation of the rectangular form of a complex number is

$$a + jb$$

where a is the real part and b is the imaginary part. The following
list shows the values of a and b for the four points in Fig. 18:

Point	a	b
1	6	3
2	−1	5
3	−7	−6
4	3	−3

19. Rectangular Form of Complex Numbers

The basic operations of arithmetic (addition, subtraction, multiplication and division) may be performed on complex numbers in
the rectangular form. These operations can be done purely by
arithmetic without regard to the geometry involved.

To add or subtract 2 (or more) complex numbers, the real parts
are added or subtracted and then the imaginary parts are added or
subtracted.

Example 22. Add $6 + j3$ and $-1 + j5$

Solution. The addition of the real parts gives $6 - 1 = 5$. The addition of the
imaginary parts gives $3 + 5 = 8$. Hence

$$6 + j3 + (-1 + j5) = 5 + j8$$

Example 23. Add $-7 - j6$ and $3 - j3$

Solution. $-7 - j6 + 3 - j3 = -4 - j9$

Example 24. Subtract $4 + j7$ from $6 + j3$

Solution. $6 + j3 - 4 + j7 = (6 - 4) + j (3 - 7) = 2 - j4$

Example 25. Subtract $-2 - j5$ from $-7 + j1$

Solution.

$$-7 + j1 - (-2 - j5) = [-7 - (-2)] + j [1 - (-5)]$$
$$= (-7 + 2) + j (1 + 5) = -5 + j6$$

20. Geometrical Addition of Complex Numbers

The process of adding two complex numbers may be done geometrically. Let us add the complex numbers of Example 22 in this
way. As shown in Fig. 19, each point is located, and a line from the
origin to each point is drawn. An arrowhead is shown at the end
of each line, for these lines have not only magnitude (length), but

also direction. A line with magnitude and direction is called a *vector*. The geometric problem we are to perform is the problem of the addition of vectors.

Vectors are added by adding their horizontal and vertical components separately. The horizontal components of the vectors are $+6$ and -1, adding to $+5$. The vertical components are $+3$ and $+5$, totaling $+8$. The sum vector is then $5 + j8$, the vector to the point P_1.

Note that the sum vector is the diagonal of a parallelogram formed by the original. This is always true, and vectors may be added by "completing the parallelogram," *i.e.* drawing the diagonal from the origin. The dotted lines going to P_1 are the missing parts of the parallelogram. To complete a parallelogram, a line is drawn from the head of each vector equal, parallel to and in the same direction as the other vector. The diagonal is then drawn from the origin to the point of intersection of the two drawn lines.

The addition of the two complex numbers of Example 23 is shown in Fig. 19. The two lines drawn in to complete the parallelogram intersect at point P_2. A line from the origin to P_2 is the sum vector, and its coordinates, $-4 - j9$, constitute the sum of the two vectors.

To multiply two complex numbers in the rectangular form, the algebraic process of the multiplication of two binomials is used. In this process, *each* term of one number is multiplied by *each* term of the second number. Like terms are then combined.

Example 26.   Multiply $2 + j4$ by $3 + j2$

Solution.

$$\begin{array}{r} 2 + j4 \\ \times\ \underline{3 + j2} \\ 6 + j12 \\ +\ j4\ +\ j^2 8 \\ \hline 6 + j16\ +\ j^2 8 \end{array}$$

However, since $j^2 = -1$, then $+ j^2 8 = -8$. The product then becomes,

$$6 + j16 + j^2 8 = 6 + j16 - 8 = -2 + j16$$

Therefore, $(2 + j4)\ (3 + j2) = -2 + j16$

Example 27.   Multiply $-5 + j1$ by $2 - j3$

Solution.

$$\begin{array}{r} -5 + j1 \\ \times\ \underline{2 - j3} \\ -10 + j2 \\ +\ j15 - j^2 3 \\ \hline -10 + j17 + 3 \end{array} = -7 + j17$$

$$[-j^2 3 = -(-1)\ (3) = +3]$$

The process of division of complex numbers in rectangular form is somewhat more complicated. We first define the "conjugate" of a number. The conjugate of a number is the number with the sign of the j term changed. Thus by definition, the conjugate of a + jb

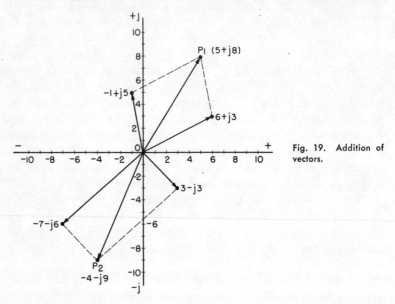

Fig. 19. Addition of vectors.

is a − jb. The conjugate of 4 − j3 is 4 + j3. The conjugate of − 5 − j7 is − 5 + j7. The sign of the real term is unchanged, but the sign of the imaginary term is reversed to make a conjugate.

Let us multiply the two general conjugate complex numbers.

$$
\begin{array}{r}
a + jb \\
\times \quad a - jb \\
\hline
a^2 + jab \\
- jab - j^2b^2 \\
\hline
a^2 \qquad\quad + b^2 \\
= \quad a^2 + b^2
\end{array}
$$

The product of two conjugate complex numbers is a *real* number equal to the sum of the squares of the a and b terms. To illustrate, let us multiply 5 − j2 by its conjugate. The product is $(5)^2 +(-2)^2 = 25 + 4 = 29$.

Multiply − 3 + j2 by its conjugate. The product is $(-3)^2 +(2)^2 = 9 + 4 = 13$.

With this technique of the conjugate we are able to divide two complex numbers. The method is broken into stepwise procedure in the following example.

Example 28.   Divide $7 + j4$ by $2 - j3$

Solution.

1. Set up the division as a fraction.

$$\frac{7 + j4}{2 - j3}$$

2. Multiply *both* numerator and denominator by the conjugate of the denominator. This is a valid operation since we are merely multiplying the fraction by 1.

$$\frac{7 + j4}{2 - j3} \times \frac{2 + j3}{2 + j3}$$

3. This leads to a new fraction, the numerator of which is the product of the two numerators and the denominator is the product of the two conjugates. We have just discussed the product of two conjugates. The new fraction is then

$$\frac{(7 + j4)\ (2 + j3)}{(2 - j3)\ (2 + j3)} = \frac{2 + j29}{13}$$

4. Now divide both the real and imaginary part of the numerator by the denominator to give the result.

$$\frac{2 + j29}{13} = \frac{2}{13} + \frac{j29}{13} = 0.154 + j2.23$$

Example 29.   Divide $- 4 - j4$ by $- 1 - j2$

Solution.

$$\frac{- 4 - j4}{- 1 - j2} \times \frac{- 1 + j2}{- 1 + j2} = \frac{(- 4 - j4)\ (- 1 + j2)}{1 + 4}$$

$$= \frac{12 - j4}{5} = 2.40 - j0.8$$

21.   The Polar Form of Complex Numbers

In addition to the rectangular form, another useful and important method of writing complex numbers is the "polar" form. In the polar form, a point in the complex plane is located by its distance from the origin and its angular displacement.*

Consider point P in Fig. 20. In rectangular notation, P is a units to the right and b units up from the origin, and is the number $a + jb$. In polar notation, P is a distance r from the origin, and

---

* A table giving the signs of trigonometric functions in the four quadrants is given on p. 92.

makes an angle Θ with the horizontal base line. In polar form, P is written as r $\angle\Theta$. This is read as "r at an angle Θ." r is called the radius vector of the point P, and is drawn with an arrowhead as

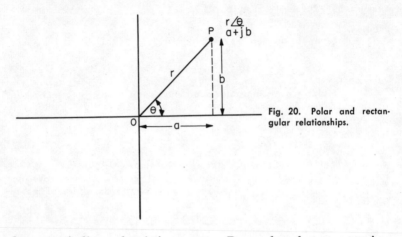

Fig. 20. Polar and rectangular relationships.

shown, to indicate that it is a vector. Remember that a vector is a quantity that has both magnitude and direction.

The relationship between the polar form and rectangular form of a complex number may be derived from Fig. 20. If the number is given as a + jb, then by trigonometry,

$$r = \sqrt{a^2 + b^2}$$

and is always taken as *positive* and

$$\tan \Theta = b/a$$

Example 30. Write the polar form of the complex number 6 + j3. Here a = 6, b = 3.

Solution.

$$r = \sqrt{6^2 + 3^2} = \sqrt{36 + 9} = \sqrt{45} = 6.71$$

$$\tan \Theta = 3/6 = 0.5$$

$$\Theta = 26.6°$$

6.71 $\angle 26.6°$ is the polar form of the number

Example 31. Write the polar form of 3 − j3.

Solution.

$$a = 3, \ b = - 3$$

$$r = \sqrt{9 + 9} = \sqrt{18} = 4.24$$

$$\tan \Theta = - 3/3 = - 1$$

Since the tangent is negative two solutions are possible, one in the second and one in the fourth quadrant. By inspection of the rectangular form it is evident that the fourth quadrant angle is desired.

$$\Theta = 360° - 45° = 315°$$

When Θ is in the fourth quadrant it may also be expressed as a negative angle, in this case $-45°$. The solution is then written in either way,

$$4.24 \ /315°$$

$$\text{or} \qquad 4.24 \ /-45°$$

If the polar form of the number is given, r/Θ, the rectangular form may be derived with the aid of Fig. 20. Again by trigonometry,

$$a = r \cos \Theta$$

$$b = r \sin \Theta$$

Example 32. Convert $5 \ /125°$ to its rectangular form.

Solution.

$$r = 5$$

$$\Theta = 125°$$

$$a = 5 \cos 125° = 5 \cos (180° - 125°)$$

$$= 5 \ (- \cos 55°) = (5) \ (- 0.574) = -2.87$$

$125°$ is a second quadrant angle, and in this quadrant the cosine is negative, while the sine is positive.

$$b = 5 \sin 125° = 5 \sin (180° - 125°) = 5 \sin 55°$$

$$= 5 \times 0.819$$

$$= 4.09$$

$$5/125° = a + jb = -2.87 + j \ 4.09$$

Example 33. Convert $4.5 \ /205°$ to the rectangular form.

$$a = 4.5 \cos 205° = 4.5 \cos (205° - 180°)$$

Solution. Remembering that in the third quadrant both sine and cosine are negative,

$$a = (4.5) \ (- \cos 25°) = (4.5) \ (- 0.906) = -4.08$$

$$b = (4.5) \ (- \sin 25°) = (4.5) \ (- 0.423) = -1.90$$

$$4.5 \ /205° = -4.08 - j1.90$$

Complex numbers in the polar form cannot be added or subtracted directly. In order to perform addition or subtraction the numbers must first be converted into rectangular form. The operations of multiplication and division are performed readily on polar quantities. In the form r $\angle\Theta$, call r the "magnitude" and Θ the angle.

To multiply two or more numbers in polar form, multiply the magnitudes and *add* the angles. Expressed mathematically,

$$(r_1 \ \angle\Theta_1) \ (r_2 \ \angle\Theta_2) = r_1 r_2 \ \angle\Theta_1 + \Theta_2$$

Example 34. Multiply the three complex numbers, $2\angle60°$, $5.85 \ \angle-22°$, $4.62 \ \angle18°$.

Solution. $\quad r_1 r_2 r_3 = 2 \times 5.85 \times 4.62 = 54.2$

$\qquad\qquad \Theta = 60° - 22° + 18° = 56°$

$\qquad\qquad 54.2 \ \angle56°$

To divide numbers in polar form, write the division as a fraction. Divide the numerator magnitude by the denominator magnitude, and *subtract* the denominator angle from the numerator angle.

$$\frac{r_1 \ \angle\Theta_1}{r_2 \ \angle\Theta_2} = \frac{r_1}{r_2} \ \angle\Theta_1 - \Theta_2$$

Example 35. Divide $7.24 \ \angle295°$ by $11.2 \ \angle140°$

Solution.

$$\frac{7.24 \quad \angle295°}{11.2 \quad \angle140°} = \frac{7.24}{11.2} \ \angle295° - 140° = 0.646 \ \angle155°$$

Example 36. Evaluate the following expression.

$$\frac{(1.62 \ \angle-74°) \ (4.35 \ \angle189°)}{2.98 \ \angle-16°}$$

Solution. Multiplying the numerator magnitudes and dividing by the denominator,

$$\frac{1.62 \times 4.35}{2.98} = 2.36$$

Adding the numerator angles and subtracting the denominator angle,

$$-74° + 189° - (-16°) = -74° + 189° + 16° = 131°$$

$$2.36 \ \angle131°$$

22. Powers and Roots of Complex Numbers

Powers and roots of complex numbers may be easily found when the numbers are in polar form. The power of a complex number is expressed by the following equation.

$$(r \ \angle\Theta)^n = r^n \ \angle n\Theta$$

The magnitude is taken to the power, but the angle is *multiplied* by the power.

Example 37. Find $(2.77 \underline{/97°})^2$

Solution. $(2.77 \underline{/97°})^2 = 2.77^2 \underline{/2 \times 97°} = 7.67 \underline{/194°}$

Example 38. Find $(4.05 \underline{/-50°})^3$

Solution. $(4.05 \underline{/-50°})^3 = 4.05^3 \underline{/3 \times -50°} = 66.2 \underline{/-150°}$
$$= 66.2 \underline{/210°}$$

Similarly for roots,

$$\sqrt[n]{r} \underline{/\Theta} = \sqrt[n]{r} \underline{\left/\frac{\Theta}{n}\right.}$$

The root is taken of the magnitude, but the angle is *divided* by the root.

Example 39. $\sqrt[3]{260} \underline{/171°}$

Solution.

$$\sqrt[3]{260} \underline{/171°} = \sqrt[3]{260} \underline{\left/\frac{171°}{3}\right.} = 6.38 \underline{/57°}$$

23. Summary of Complex Numbers

Summarizing the operations on complex numbers, addition and subtraction are possible only with the rectangular form. Powers and roots are possible only with the polar form. Multiplication and division may be done on both forms, but is almost always easier on the polar form. As a result of these facts, in our practical electrical problems where complex numbers are used, conversions from one form to the other are constantly necessary.

24. Review Questions

(1) In electrical work the "j" operator is used to indicate what operation?
(2) What are the components of a complex number?
(3) How do we add or subtract complex numbers?
(4) Define the conjugate of a complex number?
(5) What information does the polar form contain?
(6) Given two complex numbers in polar form: $25\underline{/60°}$, $5\underline{/30°}$,
    a. multiply both;
    b. divide the second into the first;
    c. take the square root of the first;
    d. square the second one.
(7) Express the answers to Question 6 in rectangular form.
(8) Given 2 numbers in rectangular form: $+6 -j5$, $-12 +j10$, use these numbers for the same four operations listed in Question 6. Express the answers in polar form.
(9) Find $(3.54 \underline{/55°})^2$.
(10) Add geometrically $-5 +j5$ and $+5 +j5$; complete the parallelogram.

Chapter 5

SERIES CIRCUITS

In Chapter 3, we studied the individual effects of resistance, inductance and capacitance in an a-c circuit. We saw that all three acted as current limiters and that the last two introduced phase shifts in the current with respect to the applied voltage. In Chapter 4, we studied the j operator and vectors. We performed arithmetic manipulations and conversions with complex numbers in both rectangular and polar forms.

25. The Properties of the Series Circuit

In this chapter we shall study the effect of combining various combinations of R, L, and C in series circuits. The following basic properties of a series circuit are listed as a review:

1. The current in any part of a series circuit is the same as in any other part. There is only one current in the circuit, and we will label it simply I.

2. By Kirchhoff's first law, the applied voltage is the sum of the individual voltage drops around the circuit. Here, the word "sum" indicates a vector sum and not a scalar or linear addition.

3. The voltage drop for any circuit element may be found by Ohm's law, the product of the current with the ohms of resistance or reactance.

26. A Series R-L Circuit

Let us first take a series circuit consisting only of resistance and inductance — R and L (Fig. 21). If an alternating voltage is ap-

plied to this series network, both R and L will exert individual in-
fluences upon the current which will result in a combined effect —
predictable from what we have already discussed.

Assuming the voltage, E, to be at a fixed frequency, the inductance will have a fixed reactance, X_L, and is so marked in the figure.
The current I is, of course, an alternating current, but for simplicity

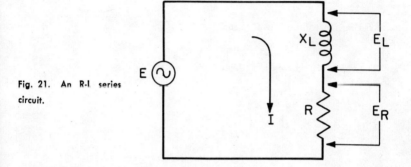

Fig. 21. An R-L series
circuit.

it is shown conventionally with a single arrow. As a result of the
current flow, there is a voltage drop across X_L marked E_L, and a
drop across R marked E_R.

The voltage relationships that exist in an R-L circuit are shown
in Fig. 22. Since I is the same throughout a series circuit, it is used
as a reference vector and is placed on the 0° line. As R does not
introduce any phase displacement in the current, then E_R must be
in phase with I and is shown as a vector on the 0° line.

An inductance, however, introduces a current lag of 90°. E_L then
must lie on the 90° or $+ j$ axis, since it is 90° *ahead* of I. By
Kirchhoff's law, the source voltage is the sum of E_L and E_R. The
sum of the two vectors is found by completing the parallelogram
and the vector E is the resultant.

E is thus seen to be a complex number. Its polar form is E $\underline{/\Theta}$,
while in rectangular form it is written $E_R + jE_L$. The two forms
are, of course equivalent, and we may write,

$$E \underline{/\Theta} = E_R + jE_L$$

where

$$E = \sqrt{E_R{}^2 + E_L{}^2}$$

and

$$\tan \Theta = E_L/E_R$$

where Θ is called the phase angle of the circuit.

Example 40. In a series R-L circuit, the voltage drops across R and L are, respectively, 30 volts and 40 volts. Find E and Θ.

Solution.

$$E = \sqrt{E_R{}^2 + E_L{}^2} = \sqrt{30^2 + 40^2} = \sqrt{2500}$$

$$= 50 \text{ volts.}$$

$$\tan \Theta = E_L \,/\, E_R = 40/30 = 1.33$$

$$\Theta = 53.1°$$

Writing the solution as a vector,

$$E = 50 \ \underline{/53.1°} \text{ volts.}$$

Therefore, the current in the circuit *lags* the voltage applied by 53.1°.

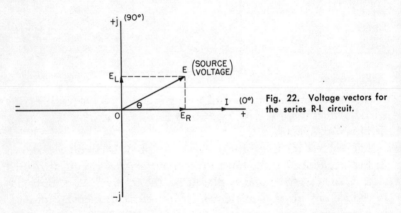

Fig. 22. Voltage vectors for the series R-L circuit.

In Fig. 21, both X_L and R contribute limiting effects on the current I. Let us define the total current-limiting effect of the circuit as the "impedance" of the circuit, and assign the symbol, Z, to the impedance. By Ohm's law,

$$I = E/Z$$

and

$$E = IZ$$

Z, R and X_L may now be related as indicated in the impedance drawing of Fig. 23. Once again I may be used as a reference vector and placed on the 0° line. By Ohm's law, the vector R is

$$R = \frac{E_R \ \underline{/0°}}{I \ \underline{/0°}} = \frac{E_R}{I} \ \underline{/0°}$$

By this analysis, R has a phase angle of 0° and is shown so.

By Ohm's law, the vector X_L may also be found.

$$X_L = \frac{E_L \: \angle 90°}{I \: \angle 0°} = \frac{E_L}{I} \: \angle 90°$$

X_L is then at 90° with respect to I, or at the $+j$ position.

By completing the parallelogram for R and X_L, we find the complex number. In the polar form it is $Z \: \angle \Theta$. In the rectangular form it is $R + jX_L$. We may then write,

$$Z \: \angle \Theta = R + jX_L$$

where $$Z = \sqrt{R^2 + X_L^2}$$

and $$\tan \Theta = X_L \: / \: R$$

In a series circuit, the phase angle Θ in both the voltage and impedance diagrams must be the same since voltage drops are proportional to ohms.

27. Power Factor

"Power factor" is defined as *the ratio of total circuit resistance to total circuit impedance*. This is R/Z in the impedance drawing.

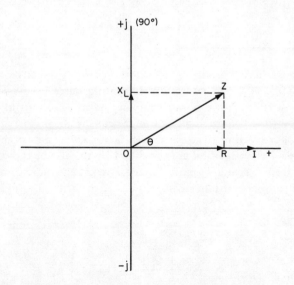

Fig. 23. Impedance vectors for the series R-L circuit.

An inspection of the figure shows that R/Z is the cosine of Θ. These relations are written as

$$\text{Power factor} = \cos \Theta = R/Z = E_R/E$$

The concept of power factor is extremely important, especially to the power company. Since it is only the resistance of a circuit that consumes power, the power factor becomes a measure of actual power used up. Recall that inductance and capacitance take power from the source during one-half cycle (positive power), but return this power to the source during the next half cycle (negative power).

Since the power company is paid only for power used, it is naturally interested in high power factors. Circuits with low power factors may have heavy currents with small power expenditure. These

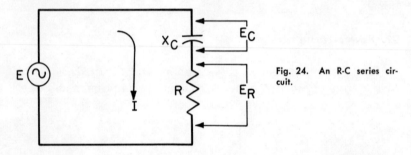

Fig. 24. An R-C series circuit.

heavy currents flow through the electric power company's lines, producing voltage drops, power losses, heating, etc. To avoid this unprofitable situation, the power company requires industrial users of electricity to maintain high power factors by means of various power factor correcting devices. Power factor correction of highly inductive machinery is often accomplished by the use of capacitors.

We may now derive a formula for the power consumed in an a-c circuit. Since only resistance consumes power, then

$$P = I^2R = I \times I \times R$$

but $$I = E/Z$$

then $$P = \frac{E \times I \times R}{Z} = EI \times R/Z$$

and $$P = EI \cos \Theta$$

This power is often called "true power" to distinguish it from "apparent power," which is simply $E \times I$ and does not take into account the phase angle between the current and voltage. True power is measured in watts, while apparent power is measured in volt-amperes. The *ratio* of the two powers is the power factor.

$$\cos \Theta = P/P_a$$

where P_a is apparent power.

Example 41. In a series circuit, R is 30 ohms and X_L is 50 ohms. The applied voltage is 115 volts. Find Z, I, Θ, the power factor, P and P_a.

Solution.

$$Z = \sqrt{R^2 + X_L^2} = \sqrt{30^2 + 50^2} = \sqrt{3400} = 58.3 \text{ ohms}$$

$$I = E/Z = 115/58.3 = 1.98 \text{ amps}$$

$$\tan \Theta = X_L/R = 50/30 = 1.67$$

$$\Theta = 59.1°$$

$$\text{Power factor} = \cos \Theta = \cos 59.1° = 0.514$$

$$P = EI \cos \Theta = 115 \times 1.98 \times 0.514 = 117 \text{ watts}$$

$$P_a = EI = 115 \times 1.98 = 228\text{-volt-amperes}$$

The correctness of the calculations can be checked by finding E_R and E_L and verifying that their vector sum is equal to E.

$$E_R = IR = 1.98 \times 30 = 59.5 \text{ volts}$$

$$E_L = IX_L = 1.98 \times 50 = 99.0 \text{ volts}$$

$$E = \sqrt{E_R^2 + E_L^2} = \sqrt{59.5^2 + 99.0^2} = \sqrt{13,340} = 115 \text{ v}$$

which checks with the given data.

28. Series R-C Circuit

A series R-C circuit presents a different set of vector relations than the R-L circuit. Figure 24 shows an R-C series circuit with the same system of labeling used in Fig. 21. Since a constant frequency is assumed, the capacitance is labeled as a fixed reactance, X_C. The voltage drops across X_C and R are marked.

Figure 25 shows the voltage vectors that exist in this circuit. Again I is taken as a reference and it lies on the 0° line. E_R is in phase with I and is also shown on the 0° line.

For the capacitance, I *leads* E_C by 90°. To maintain this relationship, E_C must be drawn on the $-j$ axis, 90° *behind* I.

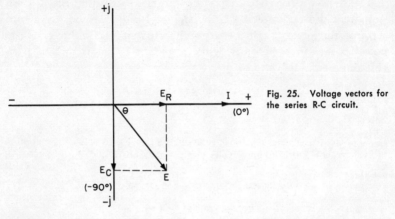

Fig. 25. Voltage vectors for the series R-C circuit.

The vector sum of E_R and E_C, as constructed by completing the parallelogram is the applied voltage E. The polar form of the complex number E is $E \angle \Theta$, while the rectangular form is $E_R - jE_C$. We can then write

$$E \angle \Theta = E_R - jE_C$$

where

$$E = \sqrt{E_R^2 + E_C^2}$$

and

$$\tan \Theta = - E_C/E_R$$

where Θ is the phase angle of the circuit.

The impedance vectors of the series R-C circuit are shown in Fig. 26. With I as the reference, R must lie on the 0° axis, while

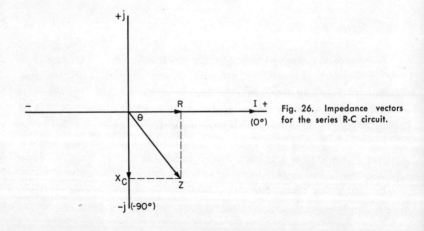

Fig. 26. Impedance vectors for the series R-C circuit.

X_C is displaced by 90° to the $-$ j axis. The impedance Z is the vector sum of R and X_C. In the polar form the impedance is Z $\underline{/\Theta}$. In the rectangular form it is $R - jX_C$. Therefore,

$$Z \underline{/\Theta} = R - jX_C$$

where
$$Z = \sqrt{R^2 + X_C^2}$$

and
$$\tan \Theta = - X_C/R$$

The concepts of power factor, true and apparent power are the same as in the R-L circuit.

Example 42. In a series R-C circuit, a current of 2 amps is measured. A voltage drop of 60 volts is read across the resistor while 45 volts is measured across the capacitor. The frequency of the generator is 800 cycles. Calculate E, Z, Θ, power factor, P, P_a, C, and R.

Solution.

$$E = \sqrt{E_R{}^2 + E_c{}^2} = \sqrt{60^2 + 45^2} = 75 \text{ volts}$$

$$Z = E/I = 75 \ / \ 2 = 37.5 \text{ ohms}$$

$$\tan \Theta = - E_C/R = - 45/60 = - 0.750$$

$$\Theta = 36.9°$$

$$\text{power factor} = \cos \Theta = \cos (- 36.9°) = 0.800$$

$$P_a = EI = 75 \times 2 = 150 \text{ volt-amps}$$

$$P = P_a \cos \Theta = 150 \times 0.800 = 120 \text{ watts}$$

$$X_C = E_C/I = 45/2 = 22.5 \text{ ohms}$$

$$C = \frac{0.159}{fX_C} = \frac{0.159}{800 \times 22.5} = 0.00000885 \text{ farad}$$

$$= 8.85 \text{ microfarads}$$

$$R = E_R/I = 60/2 = 30 \text{ ohms}$$

The calculations may be checked by verifying that Z is the vector sum of R and X_C.

$$Z = \sqrt{R^2 + X_C^2} = \sqrt{30^2 + 22.5^2}$$

$$= 37.5 \text{ ohms}$$

which checks the previously established value for Z.

29. The General Series Circuit

The general series circuit (Fig. 27) contains all three elements, R, L and C. The same convention of labeling is used as in the previous series circuits. Three possible conditions exist in such a circuit. They are

1. $X_L > X_C$. This makes $E_L > E_C$ and the circuit is inductive.*

2. $X_C > X_L$. This makes $E_C > E_L$ and the circuit is capacitive.

3. $X_L = X_C$. In this case, $E_C = E_L$, and the circuit is a "series resonant" circuit.

The series resonant circuit is discussed later in the chapter. The voltage vectors for cases 1 and 2 are sketched in Fig. 28. Let us define a voltage vector, E_X (total reactive voltage), as

$$E_X = E_L - E_C \quad \text{when } E_L > E_C$$

and

$$E_X = E_C - E_L \quad \text{when } E_C > E_L$$

Figure 28A shows the voltage vectors for $E_L > E_C$. E_X is located accordingly, and E is the vector sum of E_R and E_X. Figure 28B

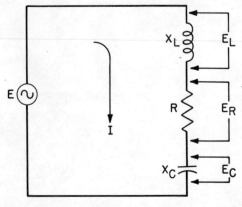

Fig. 27. The general series circuit, containing R, L and C.

shows the vectors for $E_C > E_L$. E is again the vector sum of E_R and E_X. The general equations of the circuit are,

$$E \underline{/\Theta} = E_R + jE_X \quad \text{when } E_L > E_C$$

$$E \underline{/\Theta} = E_R - jE_X \quad \text{when } E_C > E_L$$

where

$$E = \sqrt{E_R{}^2 + E_X{}^2}$$

and

$$\tan \Theta = E_X/E_R \quad \text{when } E_L > E_C$$

or

$$\tan \Theta = -E_X/E_R \quad \text{when } E_C > E_L$$

---

* The symbol $>$ is read "greater than."

Example 43. In the circuit of Fig. 27, X_L is 70 ohms, R is 8.5 ohms and X_C is 100 ohms. The current is measured at 1.25 amps. Find the individual voltage drops, E, Θ, power factor, P_a and P.

Solution.

$$E_L = IX_L = 1.25 \times 70 = 87.5 \text{ volts}$$

$$E_R = IR = 1.25 \times 8.5 = 10.6 \text{ volts}$$

$$E_C = IX_C = 1.25 \times 100 = 125 \text{ volts}$$

$$E_x = E_C - E_L = 125 - 87.5 = 37.5 \text{ volts}$$

$$E = \sqrt{E_R{}^2 + E_x{}^2} = \sqrt{10.6^2 + 37.5^2} = 39.0 \text{ volts.}$$

$$\tan \Theta = - E_x/E_R = - 37.5/10.6 = - 3.54$$

$$\Theta = - 74.2°$$

$$\text{power factor} = \cos \Theta = \cos (- 74.2°) = 0.272$$

$$P_a = EI = 39.0 \times 1.25 = 48.8 \text{ volt-amps}$$

$$P = P_a \cos \Theta = 48.8 \times 0.272 = 13.3 \text{ watts}$$

In the previous example, note that the individual voltage drops across the capacitance and the inductance were greater than the

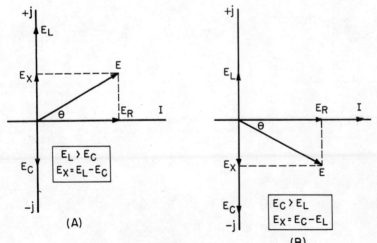

Fig. 28. Voltage vectors for the series R-L-C circuit.

applied voltage. This is perfectly valid, and violates no principles. The reason for this will become clear as we study resonance.

The applied voltage will always be greater than the net reactive voltage, E_x, and the resistive voltage, E_R, but it may be smaller, equal to or larger than the individual reactive voltages depending entirely on circuit conditions.

The impedance vectors for the series R-L-C circuit are developed in a similar manner as before. We will define the total reactance as a vector, X.

$$X = X_L - X_C \quad \text{for } X_L > X_C$$

and $$X = X_C - X_L \quad \text{for } X_C > X_L$$

Figure 29A shows the vectors for $X_L > X_C$. Z is the vector sum of R and X and Θ is a positive angle. When $X_C > X_L$, as in Fig.

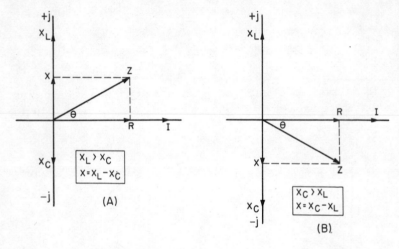

Fig. 29. Impedance vectors or the series R-L-C circuit.

29B, Θ is a negative angle. From the drawings the following relations are established,

$$Z \underline{/\Theta} = R + jX \quad \text{for } X_L > X_C$$

$$Z \underline{/\Theta} = R - jX \quad \text{for } X_C > X_L$$

where $$Z = \sqrt{R^2 + X^2}$$

and $$\tan \Theta = X/R \quad \text{for } X_L > X_C$$

or $$\tan \Theta = -X/R \quad \text{for } X_C > X_L$$

Example 44. Solve Example 43 using impedance vectors. Find Z, E, Θ and power factor.

Solution.

$$X = X_C - X_L = 100 - 70 = 30 \text{ ohms}$$

$$Z = \sqrt{R^2 + X^2} = \sqrt{8.5^2 + 30^2} = 31.2 \text{ ohms}$$

$$\tan \Theta = -X/R = -30/8.5 = -3.54$$

$$\Theta = -74.2°$$

$$\text{power factor} = \cos \Theta = \cos (-74.2°) = 0.272$$

True and apparent power of course are the same as in the previous solution.

Example 45. In a series R-L-C circuit, the following values are known. E = 15.5 volts, f = 550 Kc, L = 20 microhenries, R = 20 ohms, C = .01 microfarad. Find Z, I, Θ, power factor, and P.

Solution. It is necessary first to solve for X_L and X_C so that the circuit values will be in ohms.

$$X_L = 2\pi fL = 6.28 \times 0.550 \times 10^6 \times 20 \times 10^{-6}$$
$$= 69.0 \text{ ohms}$$

Note the conversion of kc to mc in order to make possible the cancellation of the powers of 10.

$$X_C = \frac{0.159}{fC} = \frac{0.159}{0.550 \times 10^6 \times 0.01 \times 10^{-6}}$$

$$X_C = 29.0 \text{ ohms}$$

$$X = X_L - X_C = 69 - 29 = 40 \text{ ohms}$$

$$Z = \sqrt{R^2 + X^2} = \sqrt{20^2 + 40^2} = 44.7 \text{ ohms}$$

$$I = E/Z = 15.5/44.7 = 0.347 \text{ amps} = 347 \text{ milliamps}$$

$$\tan \Theta = X/R = 40/20 = 2$$

$$\Theta = 63.4°$$

$$\text{power factor} = R/Z = 20/44.7 = 0.448 \ (= \cos 63.4°)$$

$$P = EI \cos \Theta = 15.5 \times 0.347 \times 0.448 = 2.41 \text{ watts}$$

It is interesting to note in this problem that Z was smaller than X_L, but larger than X_C and R. This is similar to the voltage situation in Example 43. We can make the statement that Z must always be greater than R and X, but may be smaller, equal to, or greater than X_L and X_C.

30. The Series-Resonant Circuit

We will now take up the case of the third possible condition that can exist in a series R-L-C circuit. This is the resonant situation where $X_L = X_C$, and therefore $E_L = E_C$. The circuit is then called a series-resonant circuit and develops certain special properties.

Figure 30 illustrates the voltage vectors that exist at resonance. Since E_C and E_L are equal and opposite then E_X is zero. In that case,

$$E = \sqrt{E_R^2 + E_X^2} = \sqrt{E_R^2}$$

or

$$E = E_R$$

Similarly,

$$\tan \Theta = E_X/E_R = 0$$

and

$$\Theta = 0°$$

The applied voltage is equal to the drop across the circuit resistance and is in phase with the current. The voltage drops

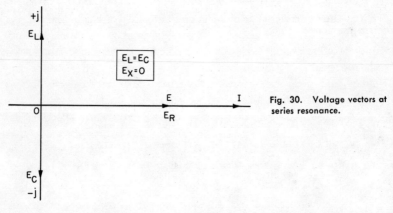

Fig. 30. Voltage vectors at series resonance.

across L and C exist (and are indeed high, as will be shown), but they cancel each other out and the net reactive voltage is zero.

Figure 31, the impedance vector drawing, shows the corresponding cancellation of X_L and X_C to give a zero X.

$$Z = \sqrt{R^2 + X^2} = \sqrt{R^2}$$

or

$$Z = R$$

and

$$\Theta = 0°$$

The impedance of a series circuit at resonance is simply the value of its resistance. This is evidently the *smallest* impedance possible for the circuit. Any value of X greater than zero must in-

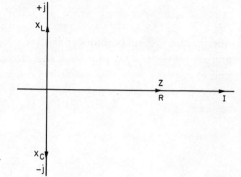

Fig. 31. Impedance vectors at series resonance.

crease the magnitude of Z. Ohm's law at resonance becomes,

$$I = E/R$$

This is the *maximum* value of current possible for the circuit.

In Fig. 32, Z and I are plotted against frequency. Starting from some frequency well below the resonant frequency, the horizontal axis represents increasing frequency through f_o, the resonant frequency and then on to well above resonant frequency. On the vertical axis, Z is in ohms and I in amperes.

Z goes from a large value at the lower frequencies, tapers off to a minimum at f_o, then rises once more. I, on the other hand, goes from small values before resonance to a maximum at resonance, then drops to low values as the frequency is increased.

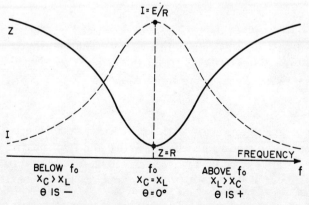

Fig. 32. Impedance and current as a function of frequency.

Below its resonant frequency, a given R-L-C circuit is capacitive since $X_C > X_L$ and Θ is a negative angle. Above its resonant frequency the circuit becomes inductive. $X_L > X_C$ and Θ is a positive angle. However, at resonance the circuit acts as a pure resistance and $\Theta = 0°$.*

We have already established the necessary conditions of series resonance as

$$X_L = X_C$$

or

$$2\pi fL = \frac{1}{2\pi fC}$$

Solving for f,

$$2\pi f^2 L = \frac{1}{2\pi C}$$

$$f^2 = \frac{1}{(2\pi)^2 LC}$$

and

$$f = \frac{1}{2\pi \sqrt{LC}} = \frac{0.159}{\sqrt{LC}}$$

The frequency at which resonance occurs for a given R-L-C circuit is thus seen to be dependent on L and C.

Another interesting phenomenon occurs at resonance. Equating E_L and E_C, and applying Ohm's law, we write

$$E_C = E_L = IX_L$$

But at resonance

$$I = E/R$$

then

$$E_C = E_L = EX_L/R$$

The ratio X_L/R is called the "Q" of a coil and is a figure of merit which describes its usefulness in electronic and electric circuits. In general, the higher the Q of a coil the better it will function. Substituting Q for X_L/R in the above equation, gives

$$E_C = E_L = QE$$

The voltage drop across each reactance at resonance is Q times the applied voltage. This effect is called the resonant rise of voltage of a series circuit. The rise in voltage becomes apparent on either side of the resonant frequency and reaches its maximum of

---

* See *Resonant Circuits*, A. Schure (ed.), (1957: John F. Rider Publisher, Inc.)

QE at resonance. This is the explanation of the situation observed in some previous problems where reactive voltages were larger than the applied voltage.

With high-Q circuits and a sizeable source voltage, quite high (and dangerous) voltages are present across the reactances, at or near resonance. For example, with a Q of 20 (a relatively low figure) and a supply voltage of 15 volts, a voltage of 300 volts will exist across both L and C at resonance. If the capacitor is not rated for this voltage it will break down.

Example 46. In a series R-L-C circuit, L is 150 microhenries, C is 250 micro-microfarads and R is 10 ohms. At what frequency will the circuit resonate? What current will flow at resonance and what will the individual voltage drops be, if the applied voltage is 2 volts?

Solution. Solving for the resonant frequency,

$$f = \frac{0.159}{\sqrt{LC}} = \frac{0.159}{\sqrt{150 \times 10^{-6} \times 250 \times 10^{-12}}}$$

$$= \frac{0.159 \times 10^9}{\sqrt{37,500}} = \frac{0.159 \times 10^9}{194}$$

$$= 0.000820 \times 10^9$$

$$= 820 \text{ kc}$$

For the current at resonance,

$$I = E/R = 2/10 = 0.2 \text{ amp}$$

For the drop across the resistor,

$$E_R = E = 2 \text{ volts}$$

For the reactive voltage drops we first must find Q.

$$Q = X_L / R = \frac{2\pi fL}{R}$$

$$Q = \frac{6.28 \times 0.820 \times 10^6 \times 150 \times 10^{-6}}{10} = 77$$

$$E_C = E_L = QE = 77 \times 2 = 154 \text{ volts}$$

31. Review Questions

(1) State Kirchhoff's first law, applied to a-c circuits.
(2) What relationship exists between voltage and current in a series circuit containing L and R?
(3) Define reactance and impedance.
(4) Define the power factor of a circuit.
(5) What is apparent power?

(6)   In a series circuit containing capacitance, E_c is shown on which axis?

(7)   What do we call a series circuit in which:

     a. $X_L = X_C$

     b. $X_C > X_L$

     c. $X_L > X_C$

     List the properties of each circuit.

(8)   What is the phase angle between E and I in a series resonant circuit?

(9)   Given L and C, what is the formula for finding the resonance frequency in a series resonant R-L-C circuit?

(10)   Why should industrial machinery have a high power factor?

Chapter 6

PARALLEL CIRCUITS AND SERIES–PARALLEL NETWORKS

The basic R-L parallel circuit is shown in Fig. 33. In order to study this circuit and other network combinations, let's review the properties of a parallel circuit:

1. The voltage across each branch of a parallel network is the same. (Therefore, we can use it as the reference vector.)

2. By Kirchhoff's second law, the total current is the sum of the individual branch currents.

3. The current in each branch is given by Ohm's law as the voltage divided by the ohms of resistance or reactance.

32. The R-L Parallel Circuit

In Fig. 33, the applied voltage E is across each branch. The branch currents are marked I_R and I_L, while line current or total current is I.

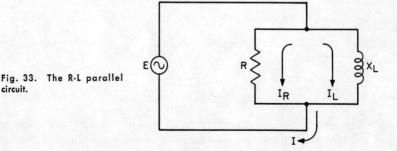

Fig. 33. The R-L parallel circuit.

The vectors used to analyze parallel networks are called current vectors. Figure 34 shows the current vectors for the simple R-L circuit. Since E is common it is used as the reference and is drawn on the 0° line. The current through R is in phase with E and also lies on the 0° line. The current through L lags E by 90°, and I_L is drawn 90° behind E or on the $-j$ axis. The line current I is the sum of the two currents and lags E by the angle Θ. Thus

$$I \underline{/\Theta} = I_R - jI_L$$

where

$$I = \sqrt{I_R{}^2 + I_L{}^2}$$

and

$$\tan \Theta = - I_L/I_R$$

$$\text{power factor} = \cos \Theta = I_R/I$$

$$Z = E/I$$

Example 47. In a circuit such as Fig. 33, R is 20 ohms and X_L is 30 ohms. If the applied voltage is 60 volts, find I, Θ, Z, power factor, P and P_a.

Solution. Solving for the branch currents,

$$I_R = E/R = 60/20 = 3 \text{ amps}$$

$$I_L = E/X_L = 60/30 = 2 \text{ amps}$$

$$I = I_R - jI_L = 3 - j2$$

$$I = \sqrt{I_R{}^2 + I_L{}^2} = \sqrt{3^2 + 2^2} = 3.60 \text{ amps}$$

$$\tan \Theta = - I_L \ / \ I_R = - 2/3 = - 0.667$$

$$\Theta = - 33.7°$$

In polar form,

$$I = 3.60 \ \underline{/- 33.7°} \text{ and lags E.}$$

$$\text{Power factor} = \cos \Theta = \cos (- 33.7°) = 0.832$$

$$P_a = EI = 60 \times 3.60 = 216 \text{ volt-amps}$$

$$P = P_a \cos \Theta = 216 \times 0.832 = 180 \text{ watts}$$

$$Z = \frac{E}{I} = \frac{E \ \underline{/0°}}{I \ \underline{/\Theta}}$$

$$= \frac{60 \ \underline{/0°}}{3.60 \ \underline{/-33.7°}}$$

$$= 16.7 \ \underline{/33.7°} \text{ ohms}$$

The impedance in this problem (Fig. 35) can be found directly from the given data. In d-c theory, the total resistance of two resistors in parallel is governed by the "product over the sum" rule. In an analogous manner, the total impedance of a two-branch parallel network is

$$Z = \frac{Z_1 Z_2}{Z_1 + Z_2}$$

Using this equation in Example 47, and remembering that inductive reactance is $+ jX_L$, we get

$$Z = \frac{(R) \ (+ jX_L)}{R + jX_L} = \frac{(R) \ (X_L \ \underline{/90°})}{R + jX_L}$$

$$= \frac{(20) \ (30) \ \underline{/90°}}{20 + j30} = \frac{600 \ \underline{/90°}}{20 + j30}$$

In order to perform the indicated division of the vectors, we convert the denominator into the polar form.

$$20 + j30 = \sqrt{20^2 + 30^2} = 36.0 \text{ for the magnitude}$$

$$\tan \Theta = 30/20 = 1.50$$

$$\Theta = 56.3° \text{ for the angle}$$

therefore, $$20 + j30 = 36.0 \ \underline{/56.3°}$$

Rewriting, $$Z = \frac{600 \ \underline{/90°}}{36.0 \ \underline{/56.3°}}$$

and $$Z = 16.7 \ \underline{/33.7°}$$

which checks with the previous solution.

33. Parallel R-L Circuit with R in Both Branches

A more complex type of parallel R-L circuit is shown in Fig. 35. This contains R as well as X_L in the inductive branch. The method

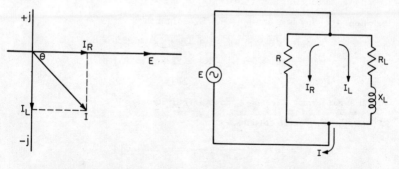

Fig. 34. The current vectors for the R-L parallel circuit.

Fig. 35. A parallel R-L circuit with R in both branches.

for solving this type of network is indicated by Fig. 36. The inductive branch, which is itself a *series* circuit, is solved for its impedance, Z_L $\underline{/\Theta_L}$. Note that the vectors for this solution in Fig. 36A use I_L as the reference since that is the common current for the inductive branch.

Once Z_L $\underline{/\Theta_L}$ is known, then by Ohm's law we can find I_L $\underline{/-\Theta_L}$. Fig. 36B shows the current-vector solution of the parallel branches

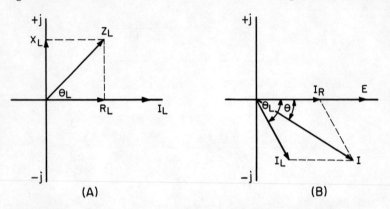

(A)   (B)

**Fig. 36.   Vectors for the solution of Fig. 35: (A) solution of the series branch;
(B) solution of the parallel branches.**

that results. I_R and I_L are added vectorially to give the line current, I $\underline{/\Theta}$. The total impedance, Z, can be found by Ohm's law or by the "product over the sum" formula.

Example 48.   In a circuit such as in Fig. 35, R = 18.5 ohms; R_L = 6.25 ohms; X_L = 22 ohms; E = 12 volts. Find the branch currents, the line current, Z, power factor, and power used.

Solution.   Solving first for the series $R_L X_L$ branch,

$$Z_L = \sqrt{R_L{}^2 + X_L{}^2} = \sqrt{6.25^2 + 22^2} = 22.9 \text{ ohms}$$

$$\tan \Theta_L = X_L \ / \ R_L = 22/6.25 = 3.52$$

$$\Theta_L = 74.1°$$

In polar form, $Z_L = 22.9 \ \underline{/74.1°} \text{ ohms}$

Finding the branch currents,

$$I_L = E/Z_L = \frac{12 \ \underline{/0°}}{22.9 \ \underline{/74.1°}} = 0.525 \ \underline{/-74.1°} \text{ amp}$$

$$I_R = E/R = \frac{12 \ \underline{/0°}}{18.5 \ \underline{/0°}} = 0.648 \ \underline{/0°} \text{ amp}$$

The line current is the vector sum of I_R and I_L. Since the branch currents are in polar form they cannot be added. Conversion to rectangular form must be performed first. Converting,

$$I_R = 0.648 \ \underline{/0°} = 0.648$$

$$I_L = 0.525 \ \underline{/-74.1°} = 0.525 \cos (-74.1°)$$

$$+ \ j0.525 \sin (-74.1°)$$

$$= 0.525 \times 0.274 + j0.525 \times (-0.962)$$

$$= 0.144 - j0.505$$

Solving for I, $I = I_R + I_L$ (vector sum)

$$I = 0.648 + 0.144 - j0.505$$

$$= 0.792 - j0.505$$

Now converting this value for I to polar form,

$$I = \sqrt{0.792^2 + 0.505^2} = 0.940 \text{ amp}$$

$$\tan \Theta = -\ 0.505/0.792 = -\ 0.638$$

$$\Theta = -\ 32.5°$$

In vector form, $\qquad I = 0.940 \ \underline{/-32.5°} \text{ amp}$

By Ohm's law,

$$Z = E/I = \frac{12 \ \underline{/0°}}{0.940 \ \underline{/-32.5°}} = 12.8 \ \underline{/32.5°} \text{ ohms}$$

The impedance could have been found by product over sum.

$$Z = \frac{RZ_L}{R + Z_L}$$

Using the polar forms for the product and the rectangular forms for the sum,

$$Z = \frac{(18.5 \ \underline{/0°}) \ (22.9 \ \underline{/74.1°})}{18.5 + j0 + 6.25 + j22}$$

$$Z = \frac{424 \ \underline{/74.1°}}{24.8 + j22}$$

Converting the denominator vector into polar form by the usual method it becomes $33.2 \ \underline{/41.6°}$. Putting this in for the denominator

$$Z = \frac{424 \ \underline{/74.1°}}{33.2 \ \underline{/41.6°}} = 12.8 \ \underline{/32.5°}$$

This checks with the solution obtained by means of currents.

$$\text{Power factor} = \cos \Theta = \cos (-32.5°)$$

$$= 0.843$$

$$P = EI \cos \Theta = 12 \times 0.940 \times 0.843$$

$$= 9.51 \text{ watts}$$

34.   A Parallel R-L Circuit with R and L in Each Branch

A still more complex circuit would have R and L in each branch
as in Fig. 37. To solve such a network, each branch would have to
be treated as a separate series circuit in the manner of Fig. 36A.

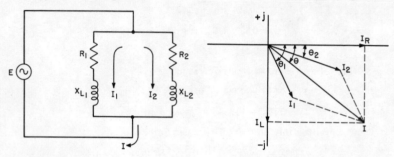

Fig. 37. R and L in each branch
of a parallel R-L circuit.

Fig. 38. Current vectors for the
parallel R-L current of Fig. 37.

The branch currents, I_1 and I_2, can then be determined and added
vectorially as in Fig. 38. An example will illustrate the technique.

Example 49. In Fig. 37, the following values exist. R_1 = 3.45 ohms; X_{L1} =
11.6 ohms R_2 = 15.3 ohms; X_{L2} = 2.75 ohms. The applied voltage is 20
volts. Find branch and line currents, circuit impedance, power factor and
power used.

Solution. Solving branch 1,

$$Z_1 = \sqrt{R_1{}^2 + X_{L1}{}^2} = \sqrt{3.45^2 + 11.6^2} = 12.1 \text{ ohms}$$

$$\tan \Theta_1 = X_{L1}/R_1 = 11.6/3.45 = 3.36$$

$$\Theta_1 = 73.4°$$

In polar form,

$$Z_1 = 12.1 \,\underline{/73.4°}$$

For current,

$$I_1 = E/Z_1 = \frac{20 \,\underline{/0°}}{12.1 \,\underline{/73.4°}} = 1.65 \,\underline{/-73.4°} \text{ amps}$$

Converting the current to the rectangular form,

$$I_1 = I_1 \cos \Theta + j \, I_1 \sin \Theta$$

$$= 1.65 \cos (-73.4°) + j1.65 \sin (-73.4°)$$

$$= 1.65 \times 0.286 - j1.65 \times 0.958$$

$$= 0.473 - j1.58$$

Branch 2 is solved similarly for Z_2 and both vector forms of I_2.

$$Z_2 = \sqrt{R_2^2 + X_{L2}^2} = \sqrt{15.3^2 + 2.75^2} = 15.6 \text{ ohms}$$

$$\tan \Theta_2 = X_{L2}/R_2 = 2.75/15.3 = 0.180$$

$$\Theta_2 = 10.2°$$

In polar form,

$$Z_2 = 15.6 \underline{/10.2°} \text{ ohms}$$

For the current,

$$I_2 = E/Z_2 = \frac{20 \underline{/0°}}{15.6 \underline{/10.2°}} = 1.28 \underline{/-10.2°} \text{ amps}$$

Converting the current to the rectangular form,

$$I_2 = 1.28 \cos (-10.2°) + j1.28 \sin (-10.2°)$$

$$= 1.28 \times 0.984 - j1.28 \times 0.177$$

$$= 1.26 - j0.226$$

The line current is found as the vector sum of the rectangular forms of the branch currents.

$$I = I_1 + I_2$$

$$= 0.473 - j1.58 + 1.26 - j0.226$$

$$= 1.73 - j1.81$$

The interpretation of this value for I is interesting. It means that the line current in the circuit has a *resistive* component of 1.73 amps and an inductive component of 1.81 amps. The resistive component is shown in Fig. 38 as I_R, and is the sum of the resistive components of I_1 and I_2. In similar fashion, the inductive component, I_L, is the sum of the inductive components of I_1 and I_2.

The line current is now converted into its polar form for its magnitude and angle of lag.

$$I = \sqrt{1.73^2 + 1.81^2} = 2.48 \text{ amps}$$

$$\tan \Theta = -\frac{1.81}{1.73} = -1.045$$

$$\Theta = -46.3°$$

In polar form,

$$I = 2.48 \underline{/-46.3°} \text{ amps}$$

The impedance is now found by Ohm's law:

$$Z = E/I = \frac{20 \underline{/0°}}{2.48 \underline{/-46.3°}} = 8.07 \underline{/46.3°}$$

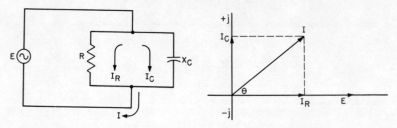

Fig. 39. The basic R-C parallel network.

Fig. 40. Current vectors for the basic R-C parallel network.

As before, the total impedance may also be found by product over sum.

$$Z = \frac{Z_1 \, Z_2}{Z_1 + Z_2}$$

Using the polar forms of the branch impedances for the multiplication in the numerator, and the rectangular forms for the addition in the denominator,

$$Z = \frac{12.1 \,\underline{/73.4°} \times 15.6 \,\underline{/10.2°}}{3.45 + j11.6 + 15.3 + j2.75}$$

$$= \frac{189 \,\underline{/83.6°}}{18.8 + j14.4}$$

The denominator is converted to polar form so that the division may be performed,

$$18.8 + j14.4 = 23.6 \,\underline{/37.5°}$$

$$Z = \frac{189 \,\underline{/83.6°}}{23.6 \,\underline{/37.5°}}$$

$$= 8.02 \,\underline{/46.1°}$$

This checks the Ohm's law calculation for Z within slide rule accuracy. We now complete the solution.

Power factor $= \cos \Theta = \cos \,(-46.3°) = 0.691$

$P_a = EI = 20 \times 2.48 = 49.6$ volt-amps

$P = P_a \cos \Theta = 49.6 \times 0.691 = 34.2$ watts

35. The Parallel R-C Circuit

The parallel R-C circuit is completely analogous to the parallel R-L circuit with the one exception that the current in a capacitive branch *leads* the voltage by the phase angle of the branch. Figure 39 illustrates the simplest form of the R-C network and the current vectors are shown in Fig. 40.

Since capacitive current I_c leads E by 90°, it is shown on the $+j$ axis. The equations are,

$$I = \sqrt{I_R^2 + I_C^2}$$

$$\tan \Theta = I_C/I_R$$

The general R-C parallel circuit is shown in Fig. 41. This has R and C in each branch and involves the type of solution encountered in the similar R-L network. As shown in Fig. 42, the branch currents, $I_1 \underline{/\Theta1}$ and $I_2 \underline{/\Theta2}$, are determined separately. The line current $I \underline{/\Theta}$ is the vector sum of the branch currents.

Example 50. In the circuit of Fig. 41, the following values exist for the circuit elements. $R_1 = 22.4$ ohms, $X_{C1} = 60.5$ ohms, $R_2 = 110$ ohms, $X_{C2} = 75.8$ ohms. $E = 150$ volts. Find branch currents, line current, impedance, power factor, apparent and true power. C_2 measures 5.25 microfarads. Find the line frequency and C_1.

Solution. We proceed first to find the branch impedances. In rectangular form,

$$Z_1 = R_1 - jX_{C1} = 22.4 - j60.5$$

$$Z_2 = R_2 - jX_{C2} = 110 - j75.8$$

Converting both to polar form,

$$Z_1 = \sqrt{R_1^2 + X_{C1}^2} = \sqrt{22.4^2 + 60.5^2} = 64.4 \text{ ohms}$$

$$\tan \Theta_1 = -X_{C1}/R_1 = -60.5/22.4 = -2.70$$

$$\Theta_1 = -69.7°$$

and,

$$Z_1 = 64.4 \underline{/-69.7°}$$

Similarly,

$$Z_2 = \sqrt{R_2^2 + X_{C2}^2} = \sqrt{110^2 + 75.8^2} = 134 \text{ ohms}$$

$$\tan \Theta_2 = -X_{C2}/R_2 = -75.8/110 = -0.690$$

$$\Theta_2 = -34.6°$$

and

$$Z_2 = 134 \underline{/-34.6°}$$

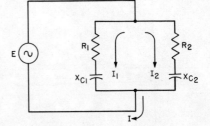

Fig. 41. A parallel R-C circuit with R and C in each branch.

The branch currents are now found by Ohm's law:

$$I_1 = \frac{E \ \underline{/0°}}{Z_1 \ \underline{/\Theta}} = \frac{150 \ \underline{/0°}}{64.4 \ \underline{/-69.7°}}$$

$$= 2.33 \ \underline{/69.7°} \ \text{amps}$$

$$I_2 = \frac{E \ \underline{/0°}}{Z_2 \ \underline{/\Theta_2}} = \frac{150 \ \underline{/0°}}{134 \ \underline{/-34.6°}} = 1.12 \ \underline{/34.6°} \ \text{amps}$$

The currents are now converted to rectangular form so that they may be added to give line current.

$$I_1 = I_1 \cos \Theta_1 + jI_1 \sin \Theta_1$$

$$= 2.33 \cos 69.7° + j2.33 \sin 69.7°$$

$$= 2.33 \times 0.347 + j2.33 \times 0.938$$

$$= 0.810 + j2.18$$

$$I_2 = I_2 \cos \Theta_2 + jI_2 \sin \Theta_2$$

$$= 1.12 \cos 34.6° + j1.12 \sin 34.6°$$

$$= 1.12 \times 0.823 + j1.12 \times 0.568$$

$$= 0.921 + j0.636$$

Finding the line current,

$$I = I_1 + I_2 = 0.810 + j2.18 + 0.921 + j0.636$$

$$= 1.73 + j2.82 \ \text{amps}$$

Converting I to polar form,

$$I = \sqrt{1.73^2 + 2.82^2} = 3.31 \ \text{amps}$$

$$\tan \Theta = 2.82/1.73 = 1.63$$

$$\Theta = 58.5°$$

$$I = 3.31 \ \underline{/58.5°} \ \text{amps}$$

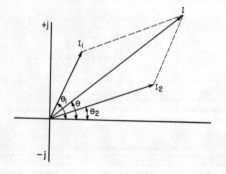

Fig. 42. Current vectors for a parallel R-C circuit with R and C in each branch.

The interpretation of the two forms of I is this. The line current has a resistive component of 1.73 amps and a capacitive component of 2.82 amps. Its magnitude (the value an ammeter would read) is 3.31 amps and it leads the applied voltage by 58.5°.

The impedance is found by Ohm's law:

$$Z = \frac{E \; \underline{/0°}}{I \; \underline{/\Theta}} = \frac{150 \; \underline{/0°}}{3.31 \; \underline{/58.5°}} = 45.4 \; \underline{/-58.5°}$$

Calculating Z by product over sum, and using the polar forms of Z_1 and Z_2 in the numerator and the rectangular forms in the denominator,

$$Z = \frac{64.4 \; \underline{/-69.7°} \times 134 \; \underline{/-34.6°}}{22.4 - j60.5 + 110 - j75.8}$$

The denominator adds to $132 - j136$. This is converted to polar form.

$$132 - j136 = 189 \; \underline{/-45.9°}$$

Rewriting the denominator in the expression for Z,

$$Z = \frac{64.4 \; \underline{/-69.7°} \times 134 \; \underline{/-34.6°}}{189 \; \underline{/-45.9°}}$$

$$= 45.5 \; \underline{/-58.4°} \text{ ohms.}$$

Thus checking the other result.

Finding power factor, P_a and P,

$$\text{power factor} = \cos \Theta = \cos(-58.5°) = 0.523$$

$$P_a = EI = 150 \times 3.31 = 496 \text{ volt-amps}$$

$$P = P_a \cos \Theta = 496 \times 0.523 = 260 \text{ watts}$$

We find the line frequency by using the data for C_2 and X_{C2}.

$$f = \frac{0.159}{X_{C2}C_2} = \frac{0.159}{75.8 \times 5.25 \times 10^{-6}} = 400 \text{ cycles}$$

Now we find C_1:

$$C_1 = \frac{0.159}{fX_{C1}} = \frac{0.159}{400 \times 60.5} = 6.58 \text{ microfarads}$$

36. The Parallel R-L-C Circuit

The simplest kind of parallel R-L-C circuit has pure R, L, and C in each branch. This is the circuit of Fig. 43. The corresponding current vectors are shown in Fig. 44. For simplicity, we may define a total reactive current, I_X.

$$I_X = I_L - I_C \text{ when } I_L > I_C$$

or

$$I_X = I_C - I_L \text{ when } I_C > I_L$$

The line current in rectangular form becomes

$$I = I_R \pm jI_X$$

For the polar form,

$$I = \sqrt{I_R{}^2 + I_X{}^2}$$

$$\tan \Theta = I_X/I_R$$

An example will illustrate the method of solution.

Example 51. In the parallel R-L-C network of Fig. 43, E is 2.5 volts at a frequency of 250 kc. If R = 12 ohms, L = 5.25 microhenries and C = 0.045 microfarads, find line current, circuit impedance and power dissipated.

Solution. X_L and X_C are first determined.

$$X_L = 6.28fL = 6.28 \times 0.250 \times 10^6 \times 5.25 \times 10^{-6}$$

$$= 8.25 \text{ ohms}$$

$$X_C = \frac{0.159}{fC} = \frac{0.159}{0.250 \times 10^6 \times 0.045 \times 10^{-6}}$$

$$= 14.1 \text{ ohms}$$

Branch currents are found.

$$I_R = \frac{E\ \underline{/0°}}{R\ \underline{/0°}} = \frac{2.5\ \underline{/0°}}{12\ \underline{/0°}}$$

$$= 0.208\ \underline{/0°} \text{ amp}$$

$$I_L = \frac{E\ \underline{/0°}}{X_L\ \underline{/90°}} = \frac{2.5\ \underline{/0°}}{8.25\ \underline{/90°}}$$

$$= 0.303\ \underline{/-90°} \text{ amp}$$

$$I_C = \frac{E\ \underline{/0°}}{X_C\ \underline{/-90°}} = \frac{2.5\ \underline{/0°}}{14.1\ \underline{/-90°}}$$

$$= 0.178\ \underline{/90°} \text{ amp}$$

The total current may now be determined by first finding the net reactive current.

$$I_X = I_L - I_C = -j0.303 + j0.178 = -j0.125 = 0.125\ \underline{/-90°} \text{ amp}$$

$$I = \sqrt{I_R{}^2 + I_X{}^2} = \sqrt{0.208^2 + 0.125^2} = 0.243 \text{ amp}$$

$$\tan \Theta = -I_X\ /\ I_R = -0.125/0.208 = -0.601$$

$$\Theta = -31.0°$$

In polar form,

$$I = 0.243\ \underline{/-31.0°} \text{ amp}$$

An interesting point is illustrated by this solution. In series circuits, the larger reactance determines the nature of the circuit. In

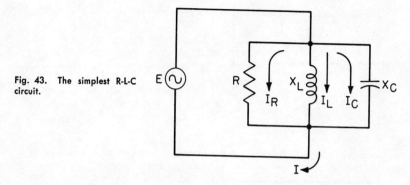

Fig. 43. The simplest R-L-C circuit.

Example 45 of the previous chapter, X_L was greater than X_C and the result was an inductive circuit. In Example 43, the circuit was capacitive as the capacitive reactance was the greater one. This is true for series circuits because the larger reactance has the larger voltage drop and controls the circuit.

In a parallel R-L-C circuit the reverse is true. The *smaller* reactance determines the nature of the circuit. In the problem we are now doing, X_L is smaller than X_C, but the line current is lagging by 31.0° and the circuit is inductive. The explanation is readily available. The *smaller* reactance has the *larger* branch current and thus controls the overall nature of the network.

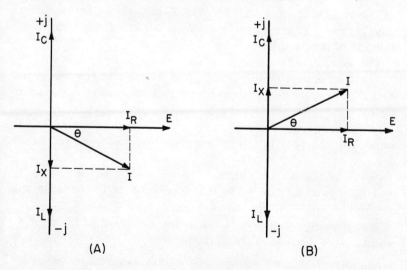

Fig. 44. Vectors for Fig. 43.

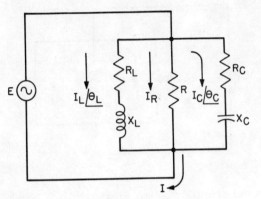

Fig. 45. The general R-L-C circuit.

Continuing the solution, we next find Z.

$$Z = \frac{E\ \underline{/0°}}{I\ \underline{/\Theta}} = \frac{2.5\ \underline{/0°}}{0.243\ \underline{/-31.0°}} = 10.3\ \underline{/31.0°}\ ohms$$

To find the power we first find power factor.

$$power\ factor = \cos \Theta = \cos (-31.0°) = 0.857$$

$$P = EI \cos \Theta = 2.5 \times 0.243 \times 0.857$$

$$= 0.521\ watt$$

37. The General R-L-C Circuit

The general R-L-C circuit contains R in both the L branch and the C branch. Figure 45 is the network, while Fig. 46 is the vector solution of the circuit.

The solution of the circuit may be outlined in four steps:

1. The branch impedances, Z_L and Z_C are calculated from the branch resistances and reactances, treating each branch as a series circuit.

2. The branch currents, $I_L\ \underline{/\Theta_L}$, $I_C\ \underline{/\Theta_C}$ and $I_R\ \underline{/0°}$ are determined by Ohm's law. These currents are shown as vectors in Fig. 46.

3. The net reactive current, $I_X\ \underline{/\Theta_X}$, is found by adding I_C and I_L. I_L and I_C must be in the rectangular vector form for the addition.

4. I, in its rectangular form, is found by the addition of I_X and I_R. It is then converted to the polar form, $I\ \underline{/\Theta}$.

Example 52. In Fig. 45, the circuit has the following constants: $R_L = 190$ ohms; $X_L = 285$ ohms; $R_C = 80$ ohms; $X_C = 215$ ohms; $R = 275$ ohms; $E = 45$ volts. Find line current, circuit impedance and power consumed.

Solution. Following the outlined procedure, we proceed to find Z_L and Z_C.

$$Z_L = \sqrt{R_L^2 + X_L^2} = \sqrt{190^2 + 285^2}$$

$$= 342 \text{ ohms}$$

$$X_L/R_L = 285/190 = 1.50$$

$$\tan 56.3° = 1.50$$

In polar form,

$$Z_L = 342 \,\underline{/56.3°} \text{ ohms}$$

$$Z_C = \sqrt{R_C^2 + X_C^2} = \sqrt{80^2 + 215^2} = 230 \text{ ohms}$$

$$- X_C/R_C = -215/80 = -2.69$$

$$\tan (-69.6°) = -2.69$$

In polar form,

$$Z_C = 230 \,\underline{/-69.6°} \text{ ohms}$$

Now, the three branch currents may be calculated:

$$I_R = \frac{E \,\underline{/0°}}{R \,\underline{/0°}} = \frac{45 \,\underline{/0°}}{275 \,\underline{/0°}} = 0.164 \,\underline{/0°} \text{ amp}$$

In rectangular form,

$$I_R = 0.164 + j0$$

$$I_L = \frac{E \,\underline{/0°}}{Z_L} = \frac{45 \,\underline{/0°}}{342 \,\underline{/56.3°}} = 0.132 \,\underline{/-56.3°} \text{ amp}$$

Converting this to rectangular form,

$$I_L = 0.132 \cos (-56.3°) + j0.132 \sin (-56.3°)$$

$$= 0.132 \times 0.555 + j0.132 \times (-0.832)$$

$$= 0.0731 - j0.110 \text{ amp}$$

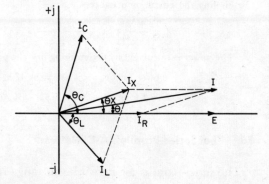

Fig. 46.   Vectors for Fig. 45.

$$I_c = \frac{E \ \underline{/0°}}{Z_c} = \frac{45 \ \underline{/0°}}{230 \ \underline{/-69.6°}} = 0.195 \ \underline{/69.6°}$$

Converting I_c to rectangular form,

$$I_c = 0.195 \cos 69.6° + j0.195 \sin 69.6°$$

$$= 0.195 \times 0.349 + j0.195 \times 0.937$$

$$= 0.0681 + j0.183 \ amp$$

The net reactive current, I_x, is found as the sum of the two reactive currents. The rectangular forms are used for the addition.

$$I_x = I_L + I_c$$

$$= 0.0731 - j0.110 + 0.0681 + j0.183$$

$$= 0.141 + j0.073 \ amp$$

The net reactive current is a capacitive current since the capacitive reactance was the smaller reactance. It has a resistive component of 0.141 amps and a capacitive component of 0.073 amps.

The line current is found as the sum of the net reactive current and resistive current.

$$I = I_R + I_x$$

$$= 0.164 + j0. + 0.141 + j0.073$$

$$= 0.305 + j0.073 \ amp$$

Converting to polar form,

$$I = \sqrt{0.305^2 + 0.073^2} = 0.314 \ amp$$

$$\tan \Theta = 0.073/0.305 = 0.239$$

$$\Theta = 13.5°$$

and

$$I = 0.314 \ \underline{/13.5°}$$

Finding the circuit impedance,

$$Z = \frac{E \ \underline{/0°}}{I \ \underline{/\Theta°}} = \frac{45 \ \underline{/0°}}{0.314 \ \underline{/13.5°}} = 143 \ \underline{/-13.5°} \ ohms$$

The power is calculated by determining first the power factor:

$$Power \ factor = \cos \Theta = \cos 13.5° = 0.972$$

$$P = EI \cos \Theta = 45 \times 0.314 \times 0.972$$
$$= 13.7 \ watts$$

38. The Series-Parallel A-C Network

The series-parallel a-c network can range from a relatively simple network, which will be considered first, to one of many extremely complicated networks. However, in spite of the complexity of the whole circuit, any series-parallel circuit can be solved by the same basic method which consists of two fundamental steps:

1. Work on each parallel network until its total impedance is expressed in rectangular form. In this way each parallel section is converted into an equivalent series section.

2. The whole circuit is now a series R-L-C network and is solved by the methods developed and illustrated in Chapter 5.

Individual problems will require more or less effort in each of the steps, but all will fall into the same pattern of solution. The first example below will illustrate the method as used to solve a relatively simple network.

Example 53. In Fig. 47 the values of the various circuit elements and the applied voltages are as shown. Find the total impedance, the line current, the power factor, true power and apparent power.

Solution. The first step is to find the impedance of the parallel network and to express it in rectangular form. This impedance, which we will call Z_p, may be determined in several ways. Two means of finding Z_p will be shown. The first is the familiar product over the sum formula. To use this formula we must first find the impedance of the $R_L X_L$ branch. Calling this branch impedance Z_L, we get,

$$Z_L = R_L + jX_L = 40 + j150$$

In polar form, this becomes

$$Z_L = 155 \; \underline{/75.1°} \text{ ohms}$$

The parallel network impedance, Z_p, by product over sum is

$$Z_p = \frac{Z_L \, X_{C1}}{Z_L + X_{C1}}$$

Using polar forms in the numerator and rectangular forms in the denominator,

$$Z_p = \frac{(155 \; \underline{/75.1°}) \; (205 \; \underline{/-90°})}{40 + j150 - j205}$$

$$= \frac{31800 \; \underline{/-14.9°}}{40 - j55}$$

Converting the denominator into polar form the equation becomes,

$$Z_p = \frac{31800 \; \underline{/-14.9°}}{68.0 \; \underline{/-53.9°}}$$

and in polar form,

$$Z_p = 468 \; \underline{/39.0°} \text{ ohms}$$

This expression is now converted to rectangular form in the usual sine and cosine manner, giving

$$Z_p = 363 + j294 \text{ ohms}$$

This has fulfilled the first step in our plan of attack. In effect, we have changed the circuit of Fig. 47A into that of Fig. 47B. The Z_p portion is shown in the dashed lines as equivalent to a resistor of 363 ohms in series with an inductive reactance of 294 ohms. The series-parallel network is now

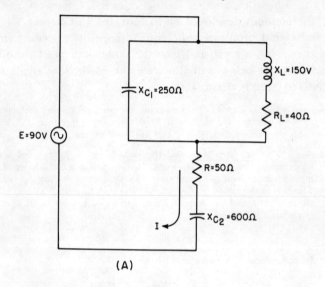

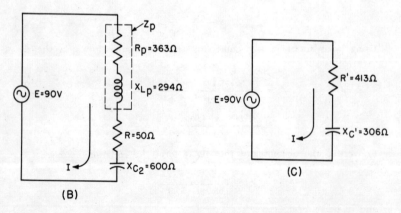

Fig. 47. The series-parallel network of Example 53 (A); its series R-L-C equivalent (B); its final equivalent circuit (C).

a series R-L-C circuit and the rest of the solution is straightforward. The total impedance, Z, is the sum of the individual resistances and reactances. We now write,

$$Z = 363 + j294 + 50 - j600$$

$$= 413 - j306 \text{ ohms}$$

Figure 47C shows this impedance as the final impedance of the original network. The entire circuit of Fig. 47A is equivalent to a resistance of 413 ohms in series with a capacitive reactance of 306 ohms.

To continue with the solution, the polar form of the impedance is now determined as

$$Z = 514 \; \underline{/-36.6°} \text{ ohms}$$

The rest of the solution follows readily. Knowing E and Z, I follows by Ohm's law.

$$I = E/Z$$

$$= \frac{90 \; \underline{/0°}}{514 \; \underline{/-36.6°}}$$

$$= 0.175 \; \underline{/36.6°} \text{ amp}$$

The current has a magnitude of 0.175 amps and leads the applied voltage by 36.6°, the phase angle of the current. By means of the phase angle the power factor is found.

$$\text{Power factor} = \cos 36.6° = 0.803$$

Real power and apparent power are found by their conventional formulas.

$$P = EI \cos \Theta$$

$$= 90 \times 0.175 \times 0.803$$

$$= 12.6 \text{ watts}$$

$$P_a = EI = 90 \times 0.175 = 15.7 \text{ volt-amps}$$

It was stated at the beginning of the example that two methods of finding the impedance of the parallel network would be shown. The first is the product over the sum as used in the actual solution. With only two branches in parallel this approach is probably as simple as any. However, with three or more parallel branches the resultant formula becomes very difficult and laborious to apply. In that case, a second method of finding the impedance of the parallel branch, which will now be given, is simpler and better.

39. Another Method of Finding the Impedance of a Parallel Branch

This method we will call the "assumed-voltage" method. To understand it we must recognize the fact that the impedance of a network depends *only* on the resistances and reactances which compose the network. The voltage across the network, and the branch currents, have no effect on the impedance. (Of course, if any element were to be damaged by the voltage or the current the impedance would change. However, this has nothing to do with our discussion.)

In the "assumed-voltage" method of determining impedance, we do just what the name implies. We assume *any* convenient voltage across our parallel circuit. This voltage has nothing to do with the applied voltage or any other voltage that may be given as part of the problem. We will use it *only* to calculate the impedance and then we will discard it. Any other information derived from this voltage, such as branch currents, is also discarded. *Only* the impedance is kept, because it is completely independent of the voltage. Regardless of the voltage assumed, the impedance will always come out the same.

By convenient voltage is meant a voltage that will simplify the arithmetic or facilitate the determination of decimal point locations. This is simply a matter of judgment.

Let us return to Example 53 and let us find Z_p by the assumed voltage method. Referring to Fig. 47A, let us assume a voltage of 205 volts across the parallel circuit. This choice eliminates one current calculation and simplifies the location of the decimal point in the other. On the basis of this assumed voltage we will find the current through X_{C1}, call it I_C, and the current through the $R_L X_L$ branch, I_L. It must be emphasized that these currents have no real existence and are only tools to help us find Z_p. They are discarded when Z_p is found. I_C and I_L are,

$$I_C = \frac{205 \text{ } \underline{/0°}}{X_{C1}} = \frac{205 \text{ } \underline{/0°}}{205 \text{ } \underline{/-90°}}$$

$$= 1 \text{ } \underline{/90°} \text{ amp} = j1 \text{ amp}$$

$$I_L = \frac{205 \text{ } \underline{/0°}}{Z_L} = \frac{205 \text{ } \underline{/0°}}{155 \text{ } \underline{/75.1°}}$$

$$= 1.32 \text{ } \underline{/-75.1°} \text{ amps}$$

I_L is converted to rectangular form, giving

$$I_L = 0.339 - j1.27$$

Let us call the assumed total current through the parallel network produced by the assumed voltage I_{as}. This will be the sum of the two assumed branch currents found above. We then write,

$$I_{as} = I_L + I_C = 0.339 - j1.27 + j1$$
$$= 0.339 - j0.270 \text{ amp}$$

This current is converted to polar form and is

$$I_{as} = 0.433 \; \underline{/-38.6°} \text{ amp}$$

The impedance of the parallel network is now found by Ohm's law, using the assumed voltage and the current that results from that assumption.

$$Z_o = \frac{205 \; \underline{/0°}}{0.433 \; \underline{/-38.6°}} = 466 \; \underline{/38.6°} \text{ ohms}$$

This is within slide-rule accuracy of the value for Z_p obtained by the previous method.

Let us apply these techniques to a problem of greater complexity, such as in the circuit of Fig. 48A. Here a three-branch parallel network is in series with resistance and inductive reactance.

Example 54.   Find Z, I, power factor, true and apparent power in the circuit of Fig. 48A.

Solution.   First, we find the branch impedances in the capacitive and inductive branches of the parallel network.

$$Z_C = R_C - jX_C = 5.85 - j27.8 \text{ ohms}$$

Converting to polar form,

$$Z_C = 28.4 \; \underline{/-78.1°} \text{ ohms}$$

Similarly for Z_L,

$$Z_L = R_L + jX_L = 24.2 + j60.8 \text{ ohms}$$

$$= 65.3 \; \underline{/68.3°} \text{ ohms}$$

With the branch impedances found, we now assume a voltage across the parallel network and proceed to calculate branch currents based on this voltage. A convenient voltage to assume is usually one equal numerically to the highest branch impedance (not including the purely resistive branch). In this case then let us assume a voltage of 65.3 volts. Solving for the branch currents,

$$I_L = \frac{65.3 \; \underline{/0°}}{Z_L} = \frac{65.3 \; \underline{/0°}}{65.3 \; \underline{/68.3°}}$$

$$= 1 \; \underline{/-68.3°} \text{ amp}$$

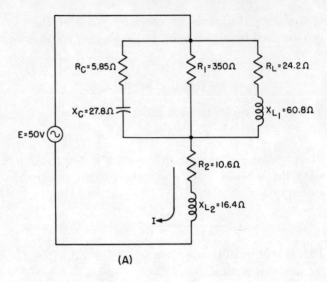

(A)

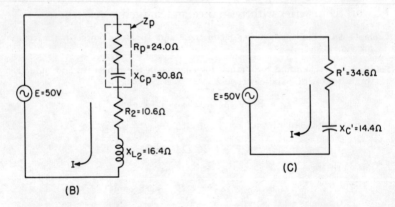

(B)

(C)

Fig. 48. The series-parallel circuit of Example 54 (A); its series R-L-C equivalent (B); its final equivalent circuit (C).

In rectangular form this becomes,

$$I_L = 0.370 - j0.929 \text{ amp}$$

In the same way,

$$I_C = \frac{65.3 \ \underline{/0°}}{Z_C} = \frac{65.3 \ \underline{/0°}}{28.4 \ \underline{/-78.1°}}$$

$$= 2.30 \ \underline{/78.1°} \text{ amps}$$

Expressed in rectangular form,

$$I_C = 0.474 + j2.25 \text{ amps}$$

The current through R_1 is found,

$$I_{R1} = \frac{65.3}{350} = 0.187 \text{ amp}$$

The total (assumed)) current is now the sum of the branch currents in their rectangular form. Calling this current I_{as} we get

$$I_{as} = I_L + I_C + IR_1$$
$$= 0.370 - j0.929 + 0.474 + j2.25 + 0.187$$
$$= 1.03 + j1.32 \text{ amps}$$

Since the numbers in the vector expression of I_{as} are small simple numbers we may solve for Z without going through the polar form. This involves a division in rectangular form and is performed by the use of conjugates. This method is explained in detail in Chapter 4; the reader can refer to that explanation should he have any doubt about the procedure.
Setting up Ohm's law for Z_p as the assumed voltage divided by I_{as} we get,

$$Z_p = \frac{65.3}{1.03 + j1.32}$$

This division is performed by multiplying both numerator and denominator by the conjugate of the denominator. When conjugates are multiplied the result is the sum of the squares of the denominator. Doing this we get,

$$Z_p = \frac{65.3}{1.03 + j1.32} \times \frac{1.03 - j1.32}{1.03 - j1.32}$$
$$= \frac{67.3 - j86.1}{1.06 + 1.74} = \frac{67.3 - j86.1}{2.80}$$
$$= 24.0 - j30.8 \text{ ohms}$$

This gives us Z_p directly in the useable rectangular form. The circuit is now a series circuit as shown in Fig. 48B. We can arrive at the total impedance by summation:

$$Z = 24.0 - j30.8 + 10.6 + j16.4$$
$$= 34.6 - j14.4 \text{ ohms}$$

In polar form this becomes,

$$Z = 37.4 \; \underline{/-22.6°} \text{ ohms}$$

The final equivalent circuit of the original series-parallel combination is illustrated in Fig. 48C. It is an R-C circuit with the values shown.

Finding the line current is next.

$$I = \frac{E}{Z} = \frac{50 \; \underline{/0°}}{37.4 \; \underline{/-22.6°}}$$
$$= 1.34 \; \underline{/22.6°} \text{ amps}$$

Now calculate power factor, true power and apparent power:

$$\text{Power factor} = \cos\Theta = \cos 22.6° = 0.923$$

$$P_a = EI = 50 \times 1.34 = 67.0 \text{ volt-amps}$$

$$P = EI\cos\Theta = 67.0 \times 0.923 = 61.9 \text{ watts}$$

When two or more parallel sections are present in the circuit, each is independently solved for its impedance. This is done in the next example which involves the circuit of Fig. 49A.

Example 55. In the circuit of Fig. 49A, find the impedance, the applied voltage, the power factor, true power and apparent power.

Solution. We first determine the parallel impedance, Z_{p1} of the top parallel network, labelled "Section 1." The branch impedances, Z_{C1} and Z_{L1}, are calculated to start.

$$\begin{aligned} Z_{C1} &= R_{C1} - jX_{C1} \\ &= 10.8 - j71.5 \text{ ohms} \end{aligned}$$

$$\begin{aligned} Z_{L1} &= R_{L1} + jX_{L1} \\ &= 104 + j55.7 \text{ ohms} \end{aligned}$$

Instead of converting Z_{C1} and Z_{L1} into polar form as in past problems, we will keep them in rectangular form and use the method of conjugates as illustrated in the previous example. To simplify the calculations we will assume a voltage of 100 volts across Section 1. Using this voltage we find the assumed branch currents. Call the branch currents I_{C1}, I_{L1} and I_{R1}.

$$\begin{aligned} I_{C1} &= \frac{100}{Z_{C1}} = \frac{100}{10.8 - j71.5} \times \frac{10.8 + j71.5}{10.8 + j71.5} \\ &= \frac{1080 + j7150}{10.8^2 + 71.5^2} = \frac{1080 + j7150}{5230} \\ &= 0.207 + j1.37 \text{ amps} \end{aligned}$$

Also,
$$\begin{aligned} I_{L1} &= \frac{100}{Z_{L1}} = \frac{100}{104 + j55.7} \times \frac{104 - j55.7}{104 - j55.7} \\ &= \frac{10,400 - j5570}{104^2 + 55.7^2} = \frac{10,400 - j5570}{13,900} \\ &= 0.750 - j0.401 \text{ amp} \end{aligned}$$

For the current in the resistive branch,

$$I_{R1} = \frac{100}{R_1} = \frac{100}{1000} = 0.100 \text{ amp}$$

Let us call the total assumed section-1 current I_1. It is now determined as the sum of the branch currents.

$$\begin{aligned} I_1 &= I_{C1} + I_{L1} + I_{R1} \\ &= 0.207 + j1.37 + 0.750 - j0.401 + 0.100 \\ &= 1.06 + j0.969 \text{ amps} \end{aligned}$$

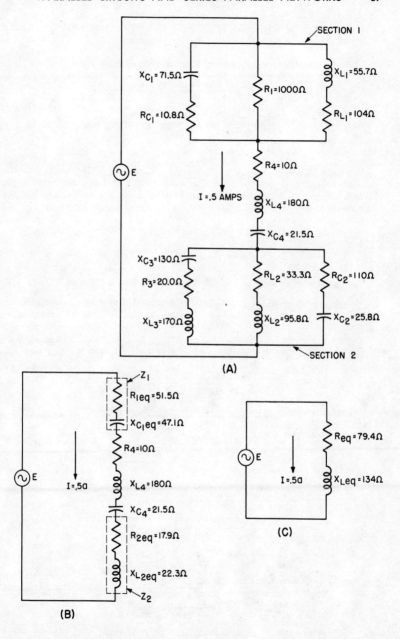

Fig. 49. The series-parallel circuit of Example 55 (A); its series equivalent (B); its final equivalent circuit (C).

We now find the impedance of Section 1 using the assumed voltage and the above current. Again we employ the conjugate method.

Calling the impedance of Section 1 Z_1 we proceed as follows,

$$Z_1 = \frac{100}{I_1} = \frac{100}{1.06 + j0.969} \times \frac{1.06 - j0.969}{1.06 - j0.969}$$

$$= \frac{106 - j96.9}{1.06^2 + 0.969^2} = \frac{106 - j96.9}{2.06}$$

$$= 51.5 - j47.1 \text{ ohms}$$

Section 1 is seen to be the equivalent of a resistance of 51.5 ohms in series with a capacitive reactance of 47.1 ohms. We will hold this result aside and proceed to a similar evaluation of Section 2. We again first find the individual branch impedances.

$$Z_{C2} = R_{C2} - jX_{C2} = 110 - j25.8 \text{ ohms}$$

and

$$Z_{L2} = R_{L2} + jX_{L2} = 33.3 + j95.8 \text{ ohms}$$

Calling the impedance of the third branch Z_a, it is written as

$$Z_a = R_3 + jX_{L3} - jX_{C3} = 20.0 + j170 - j130$$

$$= 20 + j40 \text{ ohms}$$

Let us assume a voltage of 100 volts across Section 2 and calculate the assumed branch currents. Call the branch currents I_{C2}, I_{L2} and I_a.

$$I_{C2} = \frac{100}{Z_{C2}} = \frac{100}{110 - j25.8} \times \frac{110 + j25.8}{110 + j25.8}$$

$$= \frac{11,000 + j2580}{110^2 + 25.8^2} = \frac{11,000 + j2580}{12,800}$$

$$= 0.860 + j0.202 \text{ amp}$$

For I_{L2} the equation is,

$$I_{L2} = \frac{100}{Z_{L2}} = \frac{100}{33.3 + j95.8} \times \frac{33.3 - j95.8}{33.3 - j95.8}$$

$$= \frac{3330 - j9580}{33.3^2 + 95.8^2} = \frac{3330 - j9580}{10,300}$$

$$= 0.323 - j0.930 \text{ amp}$$

And now the remaining branch current, I_a, is found.

$$I_a = \frac{100}{Z_a} = \frac{100}{20 + j40} \times \frac{20 - j40}{20 - j40}$$

$$= \frac{2000 - j4000}{20^2 + 40^2} = \frac{2000 - j4000}{2000}$$

$$= 1 - j2 \text{ amps}$$

Call I_2 the total assumed current of Section 2. It then becomes,

$$I_2 = I_a + I_{L2} + I_{C2}$$

$$= 1 - j2 + 0.323 - j0.930 + 0.860 + j0.202$$

$$= 2.18 - j2.72 \text{ amps}$$

We can now find the impedance of Section 2, which we will label Z_2.

$$Z_2 = \frac{100}{I_2} = \frac{100}{2.18 - j2.72} \times \frac{2.18 + j2.72}{2.18 + j2.72}$$

$$= \frac{218 + j272}{2.18^2 + 2.72^2}$$

$$= \frac{218 + j272}{12.2}$$

$$= 17.9 + j22.3 \text{ ohms}$$

Section 2 is thus equivalent to a resistance of 17.9 ohms in series with an inductive reactance of 22.3 ohms. Figure 49B shows the equivalent series circuit of the original circuit.

The circuit impedance, Z, is now found by addition of the resistances and reactances. Adding them in the order in which they appear in Fig. 49B, reading downward we get,

$$Z = 51.5 - j47.1 + 10 + j180 - j21.5 + 17.9 + j22.3$$

$$= 79.4 + j134 \text{ ohms}$$

The final equivalent circuit, as shown in Fig. 49C, is that of a resistance of 79.4 ohms and an inductive reactance of 134 ohms.
Z is converted into polar form to continue the solution.

$$Z = 156 \underline{/59.4°} \text{ ohms}$$

Knowing I and Z we may find the applied voltage, E. We give I the same phase angle as Z, but with the sign changed. In this way, E comes out with a zero phase angle and I shows the correct angle of lead or lag as the case may be.

$$E = IZ = 0.500 \underline{/-59.4°} \times 156 \underline{/59.4°}$$

$$= 78.0 \underline{/0°} \text{ volts}$$

The rest of the problem is readily solved.

$$\text{Power factor} = \cos 59.4° = 0.509$$

$$P_a = EI = 78 \times 0.500 = 39.0 \text{ volt-amps}$$

$$P = EI \cos \Theta = 39.0 \times 0.509 = 19.8 \text{ watts}$$

The subject of parallel resonance is a very complex one. Since it is thoroughly covered in the book *Resonant Circuits* of this series, we will not describe it now. There it is shown that at resonance, a parallel R-L-C network is at or very close to the point of maximum impedance and minimum line current.

40. Review Questions

(1) State three facts about voltage and current in a parallel circuit.

(2) Give the formula for finding the effective impedance of two impedances in parallel.

(3) Which component, E or I, leads in the capacitive branch of a parallel network?

(4) A parallel network has three branches containing R, L and C. Name the steps necessary to find the circuit impedance.

(5) Define the circuit conditions in a parallel R-L-C circuit where
   a. $X_C > X_L$;
   b. $X_L > X_C$.

(6) What general statement can be made about parallel resonance?

(7) A parallel network consists of three branches: in branch 1, R = 100 ohms; in branch 2, X_L = 50 ohms; in branch 3, X_C = 150 ohms. Find the total impedance.

(8) If a 10-volt source is applied to the network of Question 7, find the total current and the phase angle between E and I.

(9) A parallel network consists of two branches: in branch 1, 60 + j60; in branch 2, 10 − j120. A source of 100 volts is applied across the network. Find total current and effective impedance.

(10) Find the phase angle of each branch and of the total circuit in Question 9.

TABLE II
NATURAL TRIGONOMETRIC FUNCTIONS

Angle	Sine	Cosine	Tangent	Angle	Sine	Cosine	Tangent
0°	0.000	1.000	0.000	46°	.719	.695	1.036
1°	.018	1.000	.018	47°	.731	.682	1.072
2°	.035	0.999	.035	48°	.743	.669	1.111
3°	.052	.999	.052	49°	.755	.656	1.150
4°	.070	.998	.070	50°	.766	.643	1.192
5°	.087	.996	.088				
				51°	.777	.629	1.235
6°	.105	.995	.105	52°	.788	.616	1.280
7°	.122	.993	.123	53°	.799	.602	1.327
8°	.139	.990	.141	54°	.809	.588	1.376
9°	.156	.988	.158	55°	.819	.574	1.428
10°	.174	.985	.176				
				56°	.829	.559	1.483
11°	.191	.982	.194	57°	.839	.545	1.540
12°	.208	.978	.213	58°	.848	.530	1.600
13°	.225	.974	.231	59°	.857	.515	1.664
14°	.242	.970	.249	60°	.866	.500	1.732
15°	.259	.966	.268				
				61°	.875	.485	1.804
16°	.276	.961	.287	62°	.883	.470	1.881
17°	.292	.956	.306	63°	.891	.454	1.963
18°	.309	.951	.325	64°	.899	.438	2.050
19°	.326	.946	.344	65°	.906	.423	2.145
20°	.342	.940	.364				
				66°	.914	.407	2.246
21°	.358	.934	.384	67°	.921	.391	2.356
22°	.375	.927	.404	68°	.927	.375	2.475
23°	.391	.921	.425	69°	.934	.358	2.605
24°	.407	.914	.445	70°	.940	.342	2.747
25°	.423	.906	.466				
				71°	.946	.326	2.904
26°	.438	.899	.488	72°	.951	.309	3.078
27°	.454	.891	.510	73°	.956	.292	3.291
28°	.470	.883	.532	74°	.961	.276	3.487
29°	.485	.875	.554	75°	.966	.259	3.732
30°	.500	.866	.577				
				76°	.970	.242	4.011
31°	.515	.857	.601	77°	.974	.225	4.331
32°	.530	.848	.625	78°	.978	.208	4.705
33°	.545	.839	.649	79°	.982	.191	5.145
34°	.559	.829	.675	80°	.985	.174	5.671
35°	.574	.819	.700				
				81°	.988	.156	6.314
36°	.588	.809	.727	82°	.990	.139	7.115
37°	.602	.799	.754	83°	.993	.122	8.144
38°	.616	.788	.781	84°	.995	.105	9.514
39°	.629	.777	.810	85°	.996	.087	11.43
40°	.643	.766	.839				
				86°	.998	.070	14.30
41°	.656	.755	.869	87°	.999	.052	19.08
42°	.669	.743	.900	88°	.999	.035	28.64
43°	.682	.731	.933	89°	1.000	.018	57.29
44°	.695	.719	.966	90°	1.000	.000	∞
45°	.707	.707	1.000				

TABLE III
SIGNS OF TRIGONOMETRIC FUNCTIONS
IN THE FOUR QUADRANTS

Function	1st Quadrant	2nd Quadrant	3rd Quadrant	4th Quadrant
Sine	+	+	−	−
Cosine	+	−	−	+
Tangent	+	−	+	−
Cotangent	+	−	+	−

INDEX